NATIONAL AUDUBON SOCIETY® POCKET GUIDE

A Chanticleer Press Edition

Text by
John Farrand, Jr.

Edward Ricciuti, consultant

Familiar Animal Tracks

Alfred A. Knopf, New York

This is a Borzoi Book.
Published by Alfred A. Knopf, Inc.

 Published in the United States by Alfred A. Knopf, Inc., New York, and simultaneously in Canada by Random House of Canada, Limited, Toronto. Distributed by Random House, Inc., New York.

Prepared and produced by Chanticleer Press, Inc., NY.
Typeset by Stewart, Tabori & Chang, Inc., New York, and output by Pearl Pressman Liberty Communications Group, Philadelphia, Pennsylvania.
Printed and bound by Toppan Printing Co., Ltd., Hong Kong.

Published January 1993
Eleventh printing, August 2003

Library of Congress Catalog Card Number: 92-13445
ISBN: 0-679-74148-8

Contents

How to Use This Guide

Most mammals are secretive or are active only at night, and rarely allow us to catch a glimpse of them. But they often leave evidence of their presence—including tracks, droppings, burrows, and the remains of meals. Learning to read these signs is a fascinating way to learn about the hidden and little-known lives of these abundant but often elusive animals.

Coverage

This guide covers 80 of the most common mammals found in North America whose signs you would be most likely to see. Additional similar or related species are also mentioned, broadening the scope of the book.

Organization

This easy-to-use pocket guide is divided into three parts: introductory essays, illustrated accounts of the mammals and their tracks and other signs, and appendices.

Introduction

Two introductory essays explain the basics of reading the language of tracks and the traces mammals leave behind. "Finding Animal Tracks and Other Signs" tells you where to go and where to look in order to find the footprints of mammals. "Identifying Animal Tracks" offers tips on how to spot the distinctive features of tracks, and how to use droppings, trails, burrows, food caches, and foraging places of mammals to determine which species left them.

The Tracks This section includes 80 color plates of mammals, arranged according to the general similarity of their tracks. Every major group of mammals except for bats and marine mammals is represented. Opposite each color plate, along with the English and scientific names of the species or group, is a brief description of the animal's important field marks, notes on its habitat and range, and a detailed account of its tracks and important signs. An introductory paragraph provides additional information about the creature's habits and behavior. Accompanying this description are diagrams of the animal's fore print (at left), hind print (at right), typical walking track, scat, and other useful signs.

Appendices Following the species accounts, a guide to track groups shows how the species are arranged in this guide according to the similarity of their tracks. The "Guide to Mammal Orders and Families" lists the animals according to their scientific classification, the orders and families that reveal their relationships.

There is no place in North America that is without mammals. This guide is intended to help you discover and enjoy the fascinating creatures that share our continent.

Finding Animal Tracks and Other Signs

In animal-tracking, as in detective work, knowing where to look for clues is as important as actually finding them. There is no point in searching for tracks on rocks, which are too hard for animal tracks to be preserved, or in a lush meadow right after it has rained, where the grass springs back as soon as the animal has passed.

And in animal-tracking, as in detective work, one must be prepared to notice details. While tracks are often easy to see, other signs—disturbed leaf litter, scratches on the bark of a tree, or a dropping on a stone wall or log—are just as useful but easy to overlook. Instead, we must go where we know animals have been and where conditions are right for tracks and other signs to be preserved.

Tracking Surfaces

Mud, sand, dust, and snow are all surfaces on which you are likely to find the impressions left by animal footprints. Mud, along the bank of a stream or pond or around a puddle in an old dirt road, is perhaps the best place to look for tracks. Mud is often present in only small areas, but it records tracks in delicate detail. Most animals must go to water to drink, and many species, such as the Raccoon and the Mink seek their food in watery places. Still other animals, such as Muskrats, Beavers, and River Otters, are aquatic animals that leave their tracks wherever they come ashore.

In drier spots, dust and sand provide surfaces where tracks can be found. If the dust or sand is very loose, the tracks may cave in, leaving little detail, but the pattern of the tracks, and a knowledge of which animals live in the area, can help you sort the tracks out. Firmer dust or sand can contain tracks as finely preserved as those in mud. Look along the edges of roads and paths in the desert or on the plains, or in dry streambeds.

Snow is an excellent place to look for tracks. Although the quality of the tracks varies with the texture of the snow—fine, somewhat moist snow is best—you can often follow trails for long distances, observing what the animal was doing. Perhaps it was searching for food, or fleeing from a predator. Sometimes you can find the tracks of more than one animal, perhaps two of the same kind or the trails of a predator and its prey. Snow provides an ideal medium for studying the trails of different species, from the dainty, straight path left by a fox on the prowl to the irregular, ambling path of a foraging skunk.

Time of Day

No matter where you go to search for tracks, you can find them at any hour of the day. But the best time to look for tracks is early in the morning, before anything has disturbed them. Many species are most active at dawn and dusk, and if you go early enough, and move slowly and quietly, you may see the animal that made the tracks. You may even be able to follow the trail of the animal and find it. The best way to identify a track is to see for yourself the animal that has made it.

Other Signs

Signs other than tracks often require attention to detail. Did that acorn just fall from the tree, or did some animal put it there? Was that patch of flattened grass left by a resting deer? What animal left that pile of pinecone scales on a log or at the base of a tree? Many animals—skunks, the Virginia Opossum, squirrels, and armadillos—root around in leaf litter or soft soil in search of food. Study the leaf litter on the forest floor. If it has been disrupted, you know that a living creature has done the disrupting.
Other animals, among them bears, wild cats, squirrels, and Porcupines, may scratch or gnaw the bark of trees, leaving telltale signs of their presence. Many mammals, including skunks, the Mink, the River Otter, foxes, and even squirrels, leave their droppings on logs, rocks, or stone walls. And many animals have distinctive burrows.
A few of them, including moles, pocket gophers, the Pika, and jumping mice, seldom leave tracks but often leave signs that are equally diagnostic. Moles and pocket

gophers leave signs of their earthen tunnels. The Pika gathers small clusters of clipped grass stems. These little "haystacks" are the best clue to the presence of these mice, which leap over the grass rather than tunneling beneath it.

Of all the animals with which we share this planet, mammals have the most in common with humans, for we ourselves are mammals. But humans are creatures like birds, responding to sight and sound, while most other mammals rely more on their sense of smell than they do on their eyesight or hearing. They live in a sensory world that is very different from that of humans. This is a world that is largely hidden from human eyes. But as mammals go about their daily activities—searching for food, avoiding enemies, finding a mate, building safe nests, dens, or temporary resting places, or simply moving from place to place—they cannot help leaving evidence of their passing. Learning to find and understand these signs, whether they are tracks, nests, or merely disturbed leaf litter, is not difficult. Animal tracking offers a fascinating window into the hidden world of mammals. Learning to read the signs that mammals leave just takes a little detective work. Mammal-tracking is the key to the hobby of mammal-watching.

Identifying Animal Tracks

Once you have found some tracks, whether they are in mud, dust, sand, or snow, try to find the best individual track the animal has left. Look for both front and hind prints; hind prints are usually smaller. It is a good idea to make a quick sketch of the best tracks; this will not only provide you with a record of them, but will focus your attention on small details that might otherwise be overlooked.

If the animal has left a trail, follow it and note whether it is straight or crooked, purposeful or meandering. Was the animal in a hurry, as if fleeing from a predator? Or was it moving slowly, perhaps peacefully searching for food?

The trails of arboreal animals often lead from one tree to another, while those of small rodents and weasels often hug logs or stone walls, commonly appearing across openings in such walls. See if you can deduce anything about the behavior of the animal from the tracks or the trail it has left behind.

Watch for other signs—uneaten food, gnawed branches or clipped twigs, droppings, signs of an encounter between predator and prey, or evidence that there was more than one individual present. Look for nests; the tree nests of squirrels, the lodges of Beavers and Muskrats, or the burrows of chipmunks or Woodchucks will tell you right away that these animals are present, and may well be the ones that left the tracks you have found. Although many mammals are nocturnal, others are diurnal or active in the early morning, so look for the animal itself. The animal that left the tracks may be watching you while you are studying its tracks.

Finally, take note of the habitat. A knowledge of habitat can be used to eliminate many possibilities and can confirm an identification when you have made it. You will seldom find a Muskrat far from water, or a Porcupine far from trees.

Once you have garnered all the information you can, turn to the guide to track groups on pages 178–179. This will enable you to narrow the possibilities, and will sometimes lead you directly to the animal that left the tracks.

The tracks of the hoofed mammals are the easiest to identify, because only a few hoofed animals—often only one—are likely to occur in any particular area, and if more than one species is possible, hoofprints of each will be quite different. This group includes the Collared Peccary, Elk, White-tailed Deer, Mule Deer, Moose, Caribou, Pronghorn, Bison, Mountain Goat, and Bighorn Sheep.

Many animal tracks show both toe marks and a pad, like the tracks of a domestic dog or house cat. Those with five toes and a pad include the Ringtail of the Southwest and nearly all the members of the weasel family—the Marten, Fisher, Mink, Wolverine, Badger, River Otter, and the skunks.

The weasels themselves (including the Ermine) have five toes, but the fifth toe seldom leaves an impression; thus, these three species are grouped among the animals with four toes and a pad. This group also includes members of the dog family—the Coyote, Gray Wolf, and the various foxes—and members of the cat family, including the Mountain Lion, Lynx, and Bobcat.

Many tracks resemble the print of a human hand or foot. Such footlike or handlike tracks are made by bears, the Raccoon, the Beaver, the Muskrat and Nutria, the Porcupine, the Woodchuck, our two mountain-dwelling marmots, and the adaptable Virginia Opossum. All of these animals leave either a long, broad, footlike impression with well-defined toes, or a track with long "fingers." Many of our commonest mammals fall into this group.

Rabbitlike tracks are made by all the rabbits, cottontails, and hares, which leave long hind prints and much smaller fore prints. Also in this group is the Nine-banded Armadillo, which has a long hind print and a shorter fore print, but has well-defined toes, often showing signs of this animal's stout claws.

The small mammals include the rock-dwelling Pika, an alpine relative of the rabbits and hares; all the many small rodents—squirrels and chipmunks, prairie dogs, voles, mice, and rats; and the shrews. Also included here are pocket gophers and moles, subterranean mammals that seldom leave tracks and are best detected by their burrows. In the identification of tracks in this group, a knowledge of habitat and range is often a decisive factor.

Identifying animal tracks is like identifying wildflowers, birds, or any other wildlife. You may not be able to name a track the first time you come across it. But you will find it again. And the next time, you may find better tracks, or know more about the mammals in your area. Before long, you will be able to walk through the woods or across the prairie, noting the tracks and signs of mammals as easily as you would the mammals themselves. Without ever seeing a mammal you will be able to draw up a list of the animals in an area simply by spotting tracks, gnawed acorns, nests, and trails.

The Mammals

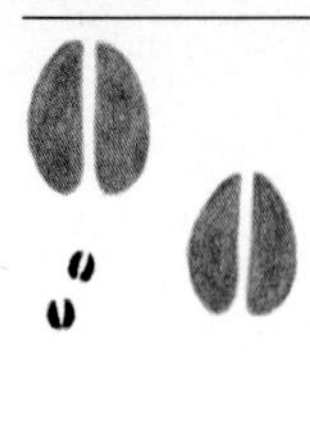

Collared Peccary *Tayassu tajacu*

These small relatives of pigs are most active early in the morning and again at dusk. Peccaries travel in bands of up to two dozen or more, foraging on cactus and other plants as well as on insects, lizards, and other small desert-dwelling animals. When a peccary is alarmed it raises the hair on its back, discharging an odor that serves as a warning signal to other members of the band.

Identification 34–37″. A small, dark, piglike animal with grizzled fur, short, slender legs, a vague, pale "collar" across shoulders, and a long, narrow snout.

Habitat and Range Deserts and dry, brushy country in S. Arizona, New Mexico, and Texas.

Signs Tracks are 1–1½″ long and rounded, usually less than 1″ apart when left by an animal walking; often seen near water; those of deer and Bighorn Sheep are longer, narrower, and more widely spaced. Droppings are looser, less like pellets than those of deer or Bighorn Sheep, and often found in caves. Peccaries frequently leave the surface of the ground disturbed by rooting. They also produce a persistent musky odor when alarmed.

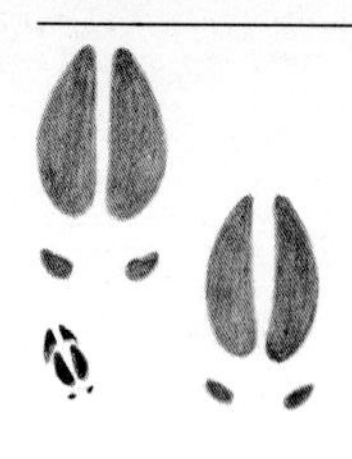

Elk *Cervus elaphus*

These large deer, also called Wapiti, spend the summer grazing on alpine meadows above the tree line. With the onset of cold weather they migrate down the slopes to sheltered valleys. Herds living in lowland areas are more sedentary. Bull Elk often have "harems" of as many as 60 cows.

Identification
7–10′ (male larger than female). A large, stocky deer, light brown or reddish brown with a pale rump patch and tail. Stag has dark neck and head, and large, spreading antlers in late summer and fall.

Habitat and Range
Forests, alpine meadows, and grasslands in W. Canada and W. United States.

Signs
Tracks are 4–4½″ long, larger and more rounded than those of deer, and more rounded than those of Moose; somewhat like more rounded prints of domestic cattle, but generally show dewclaws in moist soil or snow. Droppings are usually strong-scented pellets, 1″ or more long; in summer, droppings form "pats," smaller than those of domestic cattle. Bulls leave saplings stripped of bark after rubbing velvet from antlers, and gnaw bark from trees.

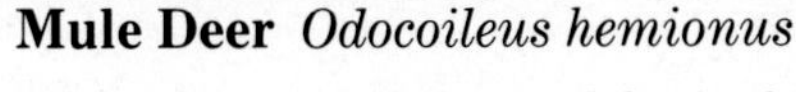

Mule Deer *Odocoileus hemionus*

This western deer, named for its large ears, forages during the day in areas where they are not molested. Mule Deer also feed at all hours in winter, when they form "yards," networks of trails in sheltered places where several gather to paw through the snow for food.

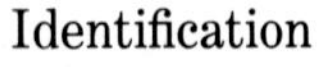

Identification 4–6½′ (male larger than female). A slender, large-eared deer with white throat and rump, white or tan underparts. Tail white with black tip. Antlers divide evenly rather than branching from curved main stem.

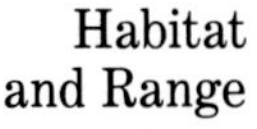

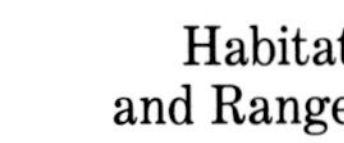

Habitat and Range Forests, woodlands, and grasslands in W. Canada and W. United States.

Signs Tracks are 2–3″ long, smaller and narrower than those of Elk or Moose, and not distinguishable from those of White-tailed Deer; dewclaws visible in moist soil; tips of hooves often widely separated. Droppings are pellets, 1″ or less long, similar to White-tailed Deer's. Often leaves beds of flattened grass where it has slept. Bucks often strip bark from saplings when rubbing velvet from antlers. Twigs bitten off by these deer lack the neat, clipped-off look of twigs taken by rabbits and rodents.

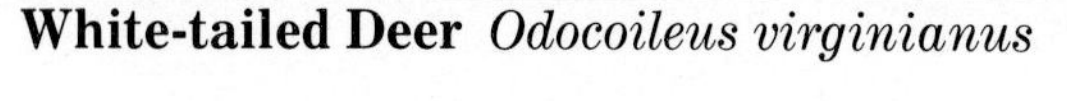

White-tailed Deer *Odocoileus virginianus*

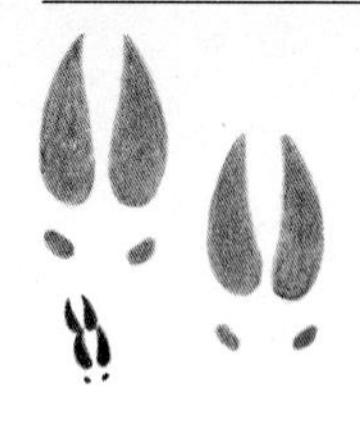

When alarmed, this abundant and familiar deer flashes the white underside of its tail and bounds away. Whitetails are much more numerous now than they were in the last century. In suburban areas, where predators are few and hunting is impossible, they have become so common that they damage woodland vegetation.

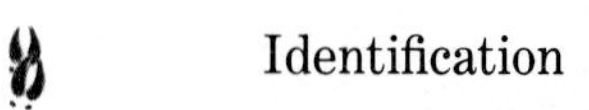

Identification: 4½–7′ (male larger than female). A slender deer, reddish brown in summer, grayer in winter, with white nose, eye ring, throat, and belly. Tail brown above, white below. Antlers have tines that branch from curved main stem.

Habitat and Range: Woodlands and suburban areas in S. Canada and United States, west to Utah, Arizona, and New Mexico.

Signs: Tracks are 2–3″ long and not distinguishable from those of Mule Deer. Droppings are pellets, 1″ or less long, similar to Mule Deer's. Often leaves beds of flattened grass where it has slept. Bucks strip bark from saplings when rubbing velvet from antlers. A browse line may be present on trees where these deer are abundant. Twigs bitten off by Whitetails lack the neat, clipped-off look of twigs taken by rabbits and rodents.

Moose *Alces alces*

The largest and the most aquatic of North American deer, the Moose spends most of its time in summer feeding on succulent plants in streams, ponds, and marshes where its tracks and droppings are most easily found. During the rutting season in the fall, bulls batter shrubs with their antlers to mark their territories.

Identification 7–10′ (male larger than female). A massive, dark brown deer with a horselike snout, a dewlap under throat, and pale legs. Stag has very large, flattened, "palmate" antlers in late summer and fall.

Habitat and Range Coniferous forests and swamps in Alaska, Canada, N. New England, N. New York, and N. Rocky Mountains.

Signs Tracks are 5–6″ long, larger and more pointed than Elk's, showing dewclaws in moist soil or snow. Droppings are large pellets, 1″ or more long, larger than those of deer; summer droppings are softer. Often leaves sleeping beds of flattened vegetation, and may produce a browse line on young conifers in winter. Browsed stems lack clean cut left by rabbits and rodents. Gnaws bark from young trees, especially aspens.

Caribou *Rangifer tarandus*

The northernmost of our deer, Caribou live in herds that sometimes number thousands of animals. In the Far North, Caribou breed on the open tundra and winter in the neighboring forest of spruce and larch, while farther to the south they stay in the forest all year. A Caribou's feet are large and its hooves are splayed, enabling it to walk on swampy tundra and ice-covered snow.

Identification 4½–8′ (male larger than female). A large, shaggy deer with long, curved antlers that are flattened at tips, with a flattened tine over forehead. Color varies from brown in south to nearly white in Far North.

Habitat and Range Coniferous forest and tundra in Alaska, Canada, and N. Idaho.

Signs Tracks are 3–5″ long, showing distinctive crescent-shaped and often widely spread hooves; dewclaws visible in soft soil or snow, and are more widely separated than those of other deer. Print of hind foot often overlaps that of larger front foot. Tracks are often abundant when left by a migrating herd. Droppings are pellets, 1″ or less long. Caribou create large pits in snow by pawing for lichens.

Pronghorn *Antilocapra americana*

This grazer is the fastest mammal in the Western Hemisphere. Except in winter, when herds may number up to 100 animals, Pronghorns travel in bands of fewer than a dozen. Living in the open on wide grasslands, foraging both day and night, and relying on their speed for safety, the animals themselves are usually much easier to find than their tracks.

Identification 4–4½′ (male larger than female). A tan or reddish-brown, long-legged, deerlike animal with bold white patches on rump, flanks, and neck; underparts white. Horns curve backward, with single short prong in males; females have shorter horns without prong.

Habitat and Range Prairies and sagebrush plains from Saskatchewan and Montana to Texas, and from California to Great Plains.

Signs Tracks are about 3″ long, narrow and pointed like those of deer, but are broader in back and never show dewclaws. Droppings are pellets, 1″ or less long, sometimes compacted into a 3″ mass, and are often deposited on bare, scraped ground. Coverings of horns that have shed can be found in late fall or winter.

Bison *Bison bison*

Only a small fraction remains of the 60 million Bison, or "Buffalo," that were the mainstay of the economy of many Indian tribes. They feed early in the morning and late in the afternoon, and spend the middle of the day resting. Bison often paw through snow to reach the grass.

Identification: 10–12½′ (male); 7–8′ (female). Unmistakable. A massive, hump-backed, large-headed animal with a shaggy, dark brown coat and short, curved horns.

Habitat and Range: Grasslands and forests in Yellowstone National Park in Wyoming, Wood Buffalo National Park in N. Canada, and in scattered public and private herds elsewhere.

Signs: Tracks are 5″ long and 5″ broad, rounded and without dewclaws. On hard ground, separate hooves may not be evident and track may resemble that of a horse. Droppings are often "pats," similar to those of domestic cattle. Produces large, dusty wallows. Often rubs body against tree trunks, sturdy fence posts, or telephone poles, leaving smooth, light-colored areas with telltale dark hairs attached, and with trampled earth below; also rubs against large boulders, which eventually become smooth.

Mountain Goat *Oreamnos americanus*

Living up high above timberline in the mountains of the West, these grazing animals avoid most predators by scrambling over the forbidding terrain, often sprinting easily across the face of what appears to be a sheer cliff. Bu the Mountain Goat is not altogether free from danger. The are not as nimble as they seem, and each year a surprising number lose their footing and plunge to their death.

Identification 4–6′ (male larger than female). A shaggy, stocky, white or off-white animal with short legs, slender black horns, black hooves, and a short beard.

Habitat and Range High mountain cliffs and rocky slopes from S. Alaska and W. Canada south to Wyoming and Colorado.

Signs Tracks are 2½–3½″ long, narrow, with hooves relatively blunt-tipped and spread in front so that track looks square Droppings are small pellets, similar to those of deer or Bighorn Sheep, but are sometimes compacted into a single large mass during summer when the diet consists of succulent plants. Often scrapes out a large, dusty bed on a sheltered ledge. Leaves conspicuous white hairs snagged on vegetation and rocks.

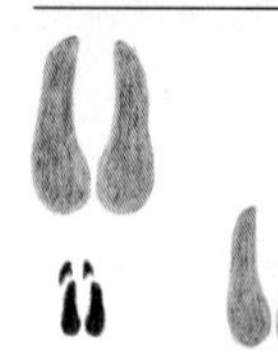

Bighorn Sheep *Ovis canadensis*

During most of the year, ewes and their offspring travel in herds of fewer than a dozen animals, but during the winter, when they are joined by the rams, groups may number as many as 100. At the outset of the fall mating season, the rams do battle for possession of the ewes, charging at one another and slamming their sturdy horns together.

Identification 4–6′ (male larger than female). A robust wild sheep with a short, smooth coat and a whitish rump. Color varies from dark brown to gray; the related Dall's Sheep (*O. dalli*) of Alaska and NW. Canada often white. Rams have massive coiled horns; ewes have slender, curved horns.

Habitat and Range Alpine meadows and rocky cliffs from SW. Canada south to Arizona, New Mexico, and W. Texas.

Signs Tracks are 3–3½″ long, narrow, with hooves relatively blunt-tipped, and sometimes spread in front so that track looks square. Droppings are small pellets, similar to those of deer or Mountain Goat's, but are sometimes compacted into a single large mass during summer, when diet consists of moist plants. Often scrapes out a large, dusty bed on a high ledge. Leaves dark hairs snagged on plants and rocks.

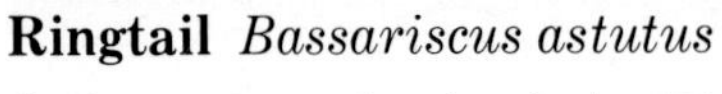

Ringtail *Bassariscus astutus*

A shy, nocturnal animal, the Ringtail, or "Cacomistle," is rarely seen and seldom leaves any sign of its presence, even in places where it is common. The ground in its desert home is usually too hard to record its footprints as it moves about under cover of darkness in search of reptiles, insects, scorpions, and fruit. It climbs easily and is slender enough to enter small crevices in rock piles, so that there are few places where its prey can hide.

Identification 24–32″. A slim, large-eared predator, with a sharp muzzle, short legs, and sandy-brown fur. Bold black and white rings on its long, bushy tail.

Habitat and Range Rocky desert country and woodlands in much of SW. United States.

Signs Tracks are 1″ long and resemble those of a house cat, except that the fore print shows an extra pad just behind the main one; most likely to be found in dust on the floor of dry, shallow caves or on sheltered ledges. Droppings are variable, but frequently contain fragments of the Ringtail's insect prey.

Marten *Martes americana*

The favorite prey of the Marten is the agile Red Squirrel, which it chases relentlessly through the trees. It also hunts on the ground, where it seeks other small mammals and birds. Fruit, nuts, and conifer seeds are an important part of its diet, and may show up in its droppings.

Identification 21–27″. A brown or yellow-brown predator, smaller and usually paler than a Fisher, with a bushy tail and a buff patch on the throat.

Habitat and Range Coniferous forests and swamps in Alaska, Canada, and border states, south in mountains to California and New Mexico.

Signs Tracks are 1½–2″ wide, slightly smaller than a Fisher's and larger than a Mink's; most frequently found in snow, spaced 6–9″ apart in a walking stride, much farther apart when an animal is bounding. Droppings are similar to those of a Mink, but may contain fragments of fruit or nuts; many droppings can be found at a "scat station," which is visited by the animal repeatedly.

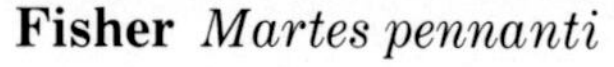

Fisher *Martes pennanti*

Larger than the related Marten, the Fisher does most of its hunting in trees. Its diet is varied, but its prey consists mainly of Snowshoe Hares and Porcupines. In taking on a Porcupine, a Fisher deftly rolls the animal over and attacks its unarmed underside. Even so, the operation is not without risk; one occasionally finds a Fisher that has died after being pierced by the quills of its intended victim.

Identification 31–42″. A slim, bushy-tailed predator, blackish or dark brown and grizzled with white on the head. It somewhat resembles a dark, overgrown squirrel.

Habitat and Range Forests in Canada and N. United States, south in mountains to California and Idaho.

Signs Tracks are 2″ wide on soil, wider on snow; larger than a Marten's or Mink's, and sometimes end abruptly at tree trunk. Droppings are variable and difficult to distinguish from Marten's, but often contain fragments of Porcupine quills. A dead Porcupine, partially eaten and lying on its back, is a good sign that a Fisher has been nearby.

Mink *Mustela vison*

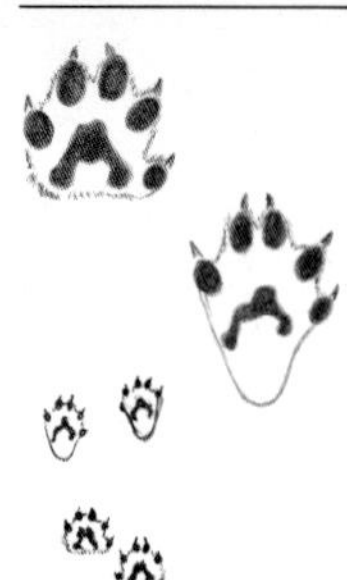

This relative of the weasels, famous for its lustrous fur, is an inhabitant of wet places, where it leaves well-defined tracks in mud. Such tracks resemble those of weasels except that the fifth toe of a weasel seldom leaves an impression. The Mink is most easily seen by paddling quietly along streams or through a swamp and watching for one bounding along the shore.

Identification — 19–28″. A dark, agile predator with glossy, brown or blackish fur and a white patch on the chin.

Habitat and Range — Shores of lakes, ponds, streams, and marshes throughout North America except in tundra of Arctic and deserts of Southwest.

Signs — Tracks are round (unless full heel mark of hind foot shows in mud), 1¼–1½″ wide, wider in snow; smaller than Marten's or Fisher's; usually found in mud at edge of water, where claw marks are apt to show. Droppings are dark and cylindrical, sometimes in segments, and may contain fur; usually placed on rocks or logs near den. Den is a 4″ hole near stream bank. Minks may leave holes in snow when they dive after prey.

Wolverine *Gulo gulo*

The "Glutton," as it was once called, is mainly a carrion-eater, but is also a predator whose strength rivals that of a bear. Even a Mountain Lion will back away if challenged at a carcass by a Wolverine. It is capable of bringing down a deer or Moose, which it will pursue with a trotting gait for days if necessary. Its jaws and large teeth are powerful enough to crush bones.

Identification 37–41″. Unmistakable. A squat, bulky predator with short legs and a short, bushy tail. Dark brown, with pale patches on head and broad, yellowish stripe down each side of body.

Habitat and Range Coniferous forest and tundra in Alaska and Canada, and south in Rocky Mountains to California and Colorado.

Signs Tracks are about 4½″ wide (larger in snow), and usually show claw marks and all 5 toes, but small toe may not print on hard soil. When only 4 toes are showing, tracks resemble Gray Wolf's, but have different shape, with extra pad behind main one. Droppings are cylindrical, up to 5″ or more in length, and may contain hair or bone fragments.

Badger *Taxidea taxus*

An expert digger armed with long, sturdy front claws, the Badger does not pass up a meal it finds on the surface, but it takes most of its prey by burrowing. It even burrows to escape other predators, rapidly scratching its way into the ground and showering dirt into the face of its pursuer.

Identification 22–30″. Unmistakable. A stocky, short-legged predator with a short, bushy tail. Grizzled buff and gray, with black and white stripes on head and black feet.

Habitat and Range Plains, grasslands, and agricultural areas in S. Canada and C. and W. United States.

Signs Tracks are 2″ long, narrow, and point inward, creating a distinctive pigeon-toed effect; claw marks are often conspicuous on fore print. Droppings are cylindrical and may contain hair or bone fragments. Den is a large, oblong hole in an area of disturbed earth, usually with bones, fur, and droppings scattered about; animal may create similar holes when burrowing after prairie dogs or ground squirrels.

River Otter *Lutra canadensis*

Few other animals leave as many signs as a playful River Otter: tracks, droppings, slides, trails, "rolling places," and even a musky odor. In places where they are not molested, the otters themselves may be fairly easy to find.

Identification 35–52″. A long, sleek, aquatic predator with a broad snout, small ears, webbed hind feet, and a long tapering tail. Brown with pale throat and breast; darker when wet.

Habitat and Range Streams and other wetlands in Alaska, Canada, and United States except for Midwest, Great Plains, and dry Southwest.

Signs Tracks are 3¼″ or more wide; claws and pad show clearly, but toes are spread and may leave no impression; webbing shows best in mud. Droppings placed on rocks or a log in water, or on a bank, usually show fish bones and scales. Slides are 12″ or more wide, in mud or snow. Often rolls on ground, leaving vegetation flattened. Sometimes leaves troughs or holes in snow when fleeing another predator.

Western Spotted Skunk *Spilogale gracilis*

This little relative of the Striped Skunk is a peaceful animal. Other animals avoid it because of its bold pattern and its striking threat display, in which it stands on its front legs, turns so that the stripes and spots on its back face the opponent, and lifts its plumelike tail. This leaves it free to wander around searching for insects, birds' eggs, small mammals, and fruit. It is less likely than its larger relative to release its strong odor.

Identification 13½–23″. A small skunk with an intricate pattern of black and white stripes and spots. Tail very bushy and often white. Very similar to Eastern Spotted Skunk (*S. putorius*) of C. and SE. United States.

Habitat and Range Deserts, mixed woodlands, open areas, and farmland in W. United States.

Signs Tracks are 1¼″ long; trail ambles or meanders even more than Striped Skunk's as the animal forages. Droppings are small, dark, and irregular. Its musky odor is not as dependable a sign as it is for the Striped Skunk.

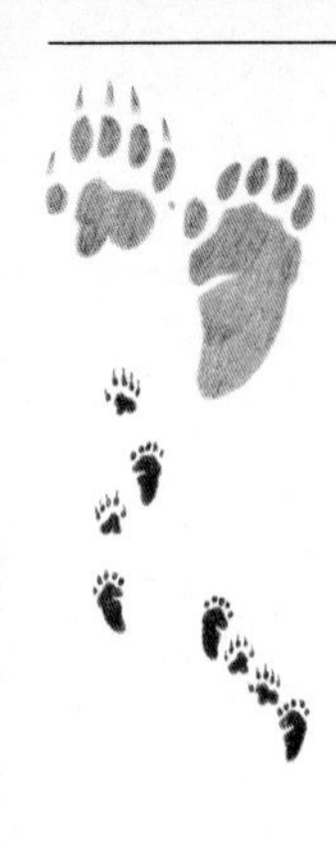

Striped Skunk *Mephitis mephitis*

Along with its tracks, which are easy to recognize, this most familiar of the skunks leaves no doubt about its presence by releasing its pungent odor, which is potent, long-lasting, and can be detected over long distances. Like other skunks, it will eat almost anything, so that its droppings are usually too variable to be identified.

Identification 22–32″. A stocky, bushy-tailed skunk, black with a white crown, a thin white line down forehead, a broad white stripe down each side, and usually with white in the tail.

Habitat and Range Forests, plains, deserts, and suburban areas in S. Canada and all of United States except parts of Southwest.

Signs Fore prints are 1–1¾″ long, hind prints 1¼–2″ long; trail usually wandering. Droppings are usually dark. Den is located in a hollow log, rocky crevice, burrow, or under a building, and is usually easily identified by strong musky odor. When foraging, skunks often leave distinctive patches of torn-up soil, dead leaves, or vegetation.

Long-tailed Weasel *Mustela frenata*

Like the related skunks, Mink, and Badger, weasels have five toes. But in weasels the fifth toe is small, so that tracks usually show just four toes. All weasels are quick and agile, with long, slim bodies that enable them to enter narrow crevices and burrows in pursuit of small mammals, reptiles, and frogs; they also prey on birds. In turn, they are captured by hawks, owls, foxes, and large snakes.

Identification 11–18″. A slender predator with short ears, short legs, and a long, black-tipped tail. Brown above and white below, usually tinged with yellow on throat and breast. In winter, some northern animals turn white except for tail tip.

Habitat and Range Woodlands, brushy areas, and farmland in S. Canada and all of United States except deserts of Southwest.

Signs Tracks are like tiny dog prints, ¾–1″ wide, usually with only 4 toes; stride varies as animal alternates between short hops and long leaps of more than 4′. Droppings are like Ermine's, but generally slightly larger. Other signs like Ermine's.

Ermine *Mustela erminea*

This species, also called the Short-tailed Weasel, is a northern relative of the Long-tailed Weasel. The tracks and other signs of the two can be difficult to tell apart, but except where their ranges overlap in the northern states, southern Canada, and the mountains of the West, geography alone will tell you which animal's tracks you are spotting.

Identification 7–13″. A slender predator with short ears and short legs. Brown above and white below, with white feet and black-tipped tail. In winter, turns all white except for black tail tip.

Habitat and Range Woodlands and thickets, usually near water, in Alaska, Canada, mountains of the West, Great Lakes Region, and NE. United States.

Signs Tracks are similar to Long-tailed Weasel's, showing only 4 of the 5 toes, but slightly smaller, ¾″ wide or less. Droppings are dark, slender, tapering at one end; often left on rocks or logs. Sometimes leaves cache of several dead mice under log, and makes holes in snow when diving after prey.

Least Weasel *Mustela nivalis*

Meadow Voles and other small rodents are the favorite food of this pint-sized predator, and its slender body allows it to follow these little animals through their tunnels durin both summer and winter. Because of its choice of prey, the Least Weasel is less likely to be found in woodlands than its larger relatives. It even makes its own nest in rodent tunnels, often after eating the original occupants. Weasels are mainly nocturnal and are seldom seen, but they are also inquisitive, and you can sometimes lure one into view by making a soft, squeaking noise.

Identification: 6–8″. A tiny weasel, brown above and white below, with no black tip on tail. In winter, most animals turn entirely white, including tail.

Habitat and Range: Open fields and brushy places in Alaska, Canada, and Midwest.

Signs: Tracks and droppings are similar to those of other weasels but much smaller. Trails are frequently found in snow along stone walls in fields or running across openings between sections of walls.

Coyote *Canis latrans*

Once a western relative of the Gray Wolf, the Coyote has expanded its range to the eastern seaboard and broadened its choice of habitat. It has now become common even in outlying suburban areas where it has adapted well to human activity.

Identification 42–52″. A buff-gray or gray wild dog with buff underparts and a black tail tip. Runs with tail pointed downward.

Habitat and Range Plains, woodlands, and thickets in Alaska, Canada, and all of United States except parts of the Southwest.

Signs Tracks are 2–2½″ wide, similar to domestic dog's but generally in a straight line; often found on hilltops or along hunting trails. Droppings similar to domestic dog's but containing hair; usually left along hunting trail and in concentrated areas not far from den. Coyotes follow regular hunting trails that wander through woods or across open terrain and may intersect. Den is a burrow in a riverbank, gully, or canyon, with entrance 1–2′ wide and with mound of scratched earth in front.

Gray Wolf *Canis lupus*

Now confined to remote areas in the North, the Gray Wolf is found in places where dogs are scarce, so that there is seldom any difficulty in identifying its tracks or other signs. Wolves prey on large mammals like deer and Moose, hunting them cooperatively in family groups or "packs" of as many as 15 animals.

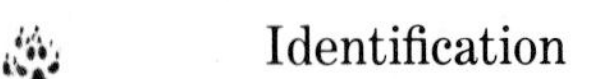

Identification: 3½–6½′. A large wild dog, usually gray grizzled with black, with black-tipped tail, but can range from black to almost pure white. Runs with tail straight out. The endangered Red Wolf (*C. rufus*), rusty or blackish, survives in parts of the SE. United States.

Habitat and Range: Forests and tundra in Alaska, Canada, and N. border of United States.

Signs: Tracks are similar to domestic dog's but much larger, up to 5″ wide; stride when walking about 30″; hind prints sometimes overlie fore prints. Droppings are like domestic dog's but larger, often containing hair. Regular hunting trails may show many tracks. Den is burrow up to 30′ deep, with entrance about 2′ wide.

Arctic Fox *Alopex lagopus*

In its bleak northern home, this little fox remains active all winter, protected from the cold by its dense fur and able to retain heat because of its short legs and ears and its compact body. It eats anything it can find, capturing prey when it has the chance, and if necessary it will eat carrion, leftovers from Polar Bear kills, and even the droppings of other animals. Life is easier during the short arctic summer, when there is an abundance of small rodents, birds, eggs, fish, crustaceans, and fruit to eat.

Identification: 30–36″. A small, short-eared fox with short legs. Bluish brown, gray, or blackish with white sides; white or pale gray in winter.

Habitat and Range: Arctic tundra and pack ice in Alaska and N. Canada.

Signs: Tracks are about 3″ wide, similar to those of Red Fox but with closely spaced toes and more rounded pads; may be blurred in winter because pads become densely furred. Droppings are small, cylindrical; may contain fragments of crab shells. Den is a shallow excavation in sandy bank with more than one entrance.

Red Fox *Vulpes vulpes*

Although they are members of the dog family, foxes leave trails that look more like those of house cats—tidy, and with the prints placed in a dainty straight line. Domestic dogs tend to be more careless; their trails wander aimlessly, reflecting the fact that they have little to fear and little prospect of capturing prey.

Identification 36–45″. A large fox with slender legs and pointed ears. Usually rich rusty color with black legs and snout, but some are blackish, blackish grizzled with white, or brown. Tail always has white tip.

Habitat and Range Woodlands, thickets, grasslands, and suburban areas in Alaska, Canada, and all of United States except Southwest Great Plains, and SE. coastal plain.

Signs Tracks are about 2″ wide, like domestic dog's but with smaller, more widely spaced toes and narrow pads; hind print more pointed than fore print. Good prints may show hair around toes and pads. In snow, tracks may be erased by tail or blurred because pads become densely furred in winter. Trail like house cat's but tracks show claws. Droppings are like dog's but vary with diet.

Kit Fox *Vulpes macrotis*

The most nocturnal of the foxes, this species and its cousin the Swift Fox use their large ears to locate prey in the dark. Their dainty trails can be found in dust or sand, and the foxes themselves can be seen darting across roads, briefly illuminated by car headlights. Like other foxes, they feed on small mammals, birds, reptiles, insects, foliage, and fruit. They are attracted by soft squeaking, but they are so stealthy that they can approach the sound, investigate it, and then slip away again without ever being noticed.

Identification: 26–34″. A small, delicate fox with very large ears. Pale gray, tinged with buff; belly white; tail tip black. The related Swift Fox (*V. velox*) of Great Plains slightly larger, with smaller ears.

Habitat and Range: Prairies, deserts, and sagebrush plains in dry W. and SW. United States.

Signs: Tracks are like those of Gray Fox, but smaller, less than 1½″ wide. Droppings are small, cylindrical. Den is burrow in ground with several entrances about 8″ wide, surrounded by earth mound and scattered bones.

Gray Fox *Urocyon cinereoargenteus*

A shyer animal than the Red Fox, this species prefers to stay in dense cover, and is less frequently seen. The fable of the fox and the grapes could never have involved a Gray Fox, a skillful climber that can easily reach fruit and has been known to escape from predators by taking refuge among the branches of trees.

Identification 31–44″. A large fox with pointed ears. Gray grizzled with black or white, rusty below and on sides of neck, and white on throat. Tail has black tip.

Habitat and Range Forests and thickets throughout United States except in parts of Great Plains and mountains of Northwest.

Signs Tracks are 1½″ wide, similar to Red Fox's but smaller, with larger toes; prints not blurred by hair in winter. Trail like house cat's but with obvious claw marks; claws are long for a fox, and claw marks may be all that shows when animal has been running. Droppings are like dog's, but vary with diet; often contain pits of berries. Den is usually a natural cavity, sometimes with snagged fox hairs or scattered bones in evidence. Leaves urine marks in snow beside trail.

Mountain Lion *Felis concolor*

Also called the Puma or Cougar, this big cat is a solitary hunter that is seldom seen. It preys mainly on deer, which it attacks from ambush, but will also take other animals, including insects. It is known to be declining throughout its range, but it is so secretive that estimates of its population are hard to obtain.

Identification 6–8′. A very large sandy or tawny cat with a long, black-tipped tail and a black patch on each side of muzzle.

Habitat and Range Forests, swamps, and deserts in W. Canada and W. United States, and in scattered places in East.

Signs Tracks are like house cat's but rounder and very large; fore print 3–4″ wide, hind print slightly smaller. Trail is a straight line. When animal is bounding, hind prints may land in fore prints, and tail may leave impression in snow. Droppings are variable, sometimes in large pellets with hair or bone; usually not buried, but surrounded by scratch marks. Leaves scratches on tree-trunk "scratching post," larger and higher than those of Lynx or Bobcat.

Lynx *Felis lynx*

A densely furred cat of cold regions, the Lynx prowls the forest in search of Snowshoe Hares. Like the hare, it too has "snowshoes," large, furry feet that enable it to stalk its prey in silence and travel fast in snow. It will take other prey, but it is so dependent upon the Snowshoe Hare that when the hare population crashes, the number of Lynxes drops as well.

Identification 29–42″. A large-footed, long-legged cat with long black ear tufts and long "sideburns." Buff or sandy grizzled with black. Stubby tail has black tip.

Habitat and Range Deep coniferous forest in Alaska, Canada, N. border states and south in Rocky Mountains to Colorado.

Signs Tracks are similar to Bobcat's but even rounder and larger; fore print about 4″ wide, hind print slightly smaller prints look even larger in snow, when furred pad also makes them blurry. Trail similar to Bobcat's, but Lynx more likely to bound, leaving widely spaced tracks. Droppings are like Bobcat's. Leaves scratches on low, tree-trunk "scratching post." Covers its kills with a thin layer of snow or litter.

Bobcat *Felis rufus*

Like the Lynx, this cat prefers rabbits and hares, but in the warmer regions there is a greater variety of animals to hunt, so Bobcats seldom have serious trouble finding live prey. When they do, they switch to carrion. This adaptability allows them to survive near humans; occasionally, they even enter suburban areas.

Identification 28–49½″. A large-footed cat with short ear tufts and short "sideburns." Buff or sandy, vaguely spotted with blackish. Stubby tail black above with white tip.

Habitat and Range Forests, swamps, deserts, and farmland in S. Canada and throughout United States except in Midwest.

Signs Tracks are round and large; fore print nearly 2″ wide, hind print slightly smaller; print like domestic dog's, but without claw marks; pad shows scalloping in front and behind (dog's is scalloped only behind). Trail a narrow line; walking animal often places hind foot on fore print; prints about 1′ apart. Droppings are like dog's but usually surrounded by scratch marks. Leaves scratches on low, tree-trunk "scratching post." Covers its prey with a loose layer of leaves or branches.

Black Bear *Ursus americanus*

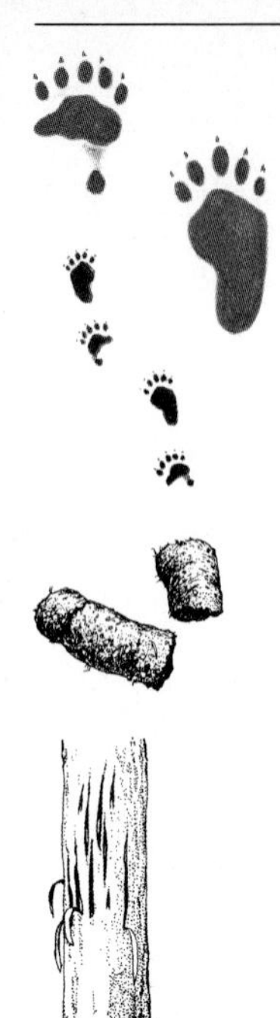

This is the most widespread of our bears, found in nearly all wooded parts of North America. Formerly rare, its population has recently increased and is now common in many areas. Like all bears, it should be treated with caution. Its diet consists mainly of fruit, insects, and small mammals, and its most conspicuous signs, apart from its tracks, are places where it has been mauling logs or digging in the ground in search of a meal.

Identification 5–6′. A medium-sized bear with a narrow head and snout. Usually black, but sometimes cinnamon-brown in West or white in coastal British Columbia.

Habitat and Range Forests and wooded swamps in Alaska, Canada, and most wooded parts of United States.

Signs Hind prints are footlike, 7–9″ long, with 5 toes (small outer toe sometimes does not show); fore prints 4″ long, 5″ wide; claw marks usually show in soil or mud; hind print usually placed ahead of fore print. Droppings are like dog's but larger; may contain hair, seeds, or plant fibers. Leaves long claw marks on standing trees, and tears open decayed logs in search of food.

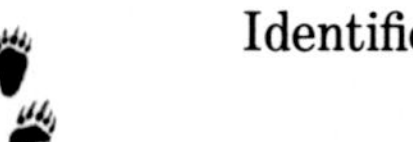

Grizzly Bear *Ursus arctos*

This massive but fast-moving predator survives only in remote areas of the West and in the Far North. It feeds on everything from berries and fish to deer and other large mammals. Like a Lynx or Bobcat, it may conceal its prey and return to it many times before moving on.

Identification — 6–8′. A large bear with a flat face and a large hump on shoulders. Variable, but usually brown, often grizzled with white.

Habitat and Range — Alpine meadows and coniferous forests in Alaska, N. and W. Canada, and south in Rocky Mountains to Wyoming.

Signs — Tracks are similar to Black Bear's, but larger (10–12″ long, 7–8″ wide), with longer claw marks in front of fore print. Often follows well-established, meandering trails. Droppings are like Black Bear's but up to 2″ in diameter. Leaves claw marks or stripped bark on standing trees, usually higher than Black Bear. Tears up decayed logs and berry bushes. Leaves large pits in ground when digging for rodents. Covers large prey with branches, leaf litter, or soil. Caution: Grizzlies usually stay close to cached prey.

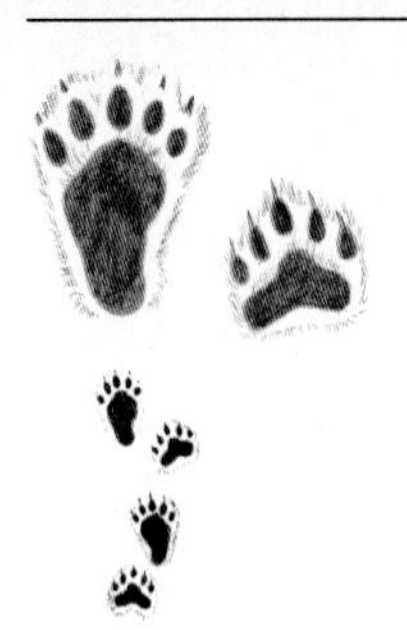

Polar Bear *Ursus maritimus*

The most formidable hunter of the Arctic, the Polar Bear preys mainly on seals, but will also feed at the carcasses of stranded whales, and will take birds, their eggs, and even fruit. Its seal kills, often found lying at the edge of the ice surrounded by the tracks of Arctic Foxes, are the most easily found of its signs. It swims well, and in pursuit of seals it often travels out of sight of land. With sharp eyes and a keen sense of smell, and aided by its white color, it can stalk a basking seal, attacking swiftly before the seal can gain the safety of the water.

Identification — 7–11′. A large, white or off-white bear with long legs and a long neck.

Habitat and Range — Arctic tundra and pack ice on coast and islands of N. Alaska and Canada.

Signs — Hind prints are 12–13″ long, 9″ or more wide, fore prints shorter; tracks often blurred by hair; claw marks seldom show. Droppings are like those of Grizzly Bear. Leaves seal kills on ice after eating blubber. The Polar Bear is the only bear that leaves signs on ice floes.

Raccoon *Procyon lotor*

This adaptable animal raises as many as seven young at a time, and so it is common wherever it occurs. Its signs are also common. Since Raccoons seek most of their food near water, the bank of a stream is a good place to look for indications of its presence. In the suburbs, where pavements are hard and lawns are well manicured, the Raccoon seldom leaves tracks and may go undetected.

Identification: 26–40″. A chunky animal with a narrow muzzle. Gray-brown or brown, with a black-and-white banded tail and a black mask bordered with white.

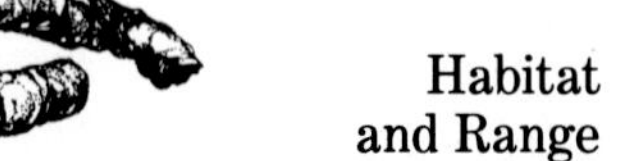

Habitat and Range: Forests, open woods, and suburban areas in S. Canada and all of United States except parts of Rocky Mountains.

Signs: Hind prints are 3¼–4¼″ long, like small human foot with long toes; fore prints 3″ long and wide, like small human hand. Places hind foot next to front foot when walking; bounds when running, leaving paired hind prints ahead of fore prints. Droppings vary with diet; often cylindrical or segmented and left on logs, rocks, or tree limbs. Den is usually in hollow tree, where bark may be scratched around the entrance.

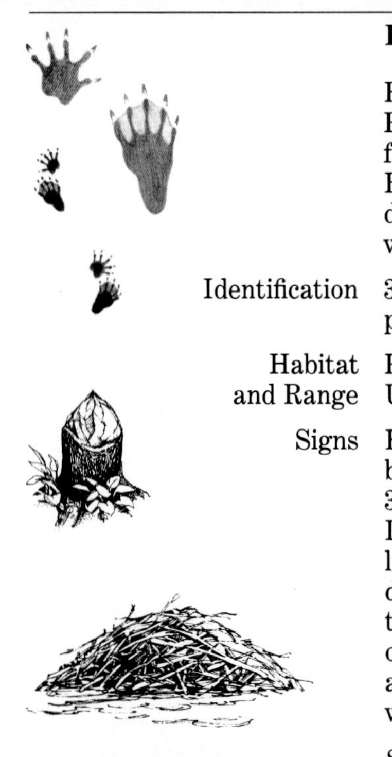

Beaver *Castor canadensis*

Few animals leave signs more obvious than those of the Beaver. By early in this century Beavers had disappeared from most of their range, but they are staging a comeback: Even in the crowded Northeast, one can now find lodges, dams, and gnawed tree stumps in places where Beavers were unknown a few decades ago.

Identification 36–47″. A large, dark brown, aquatic rodent with a broad, paddle-shaped tail and webbed hind feet.

Habitat and Range Ponds, streams, and marshes in Alaska, Canada, and all of United States except desert Southwest and Florida.

Signs Hind prints are 6″ or more long, like small human foot but with long, widespread toes joined by web; fore prints 3″ wide. Trail often partly or wholly erased by flat tail. Droppings are oval pellets, seldom seen on land. Builds lodge up to 6′ high out of sticks, reeds, and mud; builds dam of same materials across stream. Leaves pointed, tooth-marked stumps when cutting down saplings for bark or building material, and gnaws bark from aspens, willows, and birches. When alarmed, makes loud, slapping sound with tail on water.

Muskrat *Ondatra zibethicus*

This familiar inhabitant of fresh water spends most of its time gathering plant food. It can often be seen swimming across a pond with a great mass of leaves or marsh plants. While harvesting all this food, it sometimes drops some; a scattering of clipped plants floating on the surface of a pond or marsh is a sure sign that there is at least one Muskrat in residence.

Identification — 16–24½″. A large, dark brown, aquatic rodent with a hairless, vertically flattened tail and partly webbed hind feet.

Habitat and Range — Ponds, streams, and marshes in Alaska, Canada, and all of United States except desert Southwest and Florida.

Signs — Hind prints are 2–3″ long with five toes, and with webs seldom showing; fore prints shorter and smaller, usually with only four toes showing. Trail often includes drag mark of tail. Droppings are small pellets, usually in clusters on bank or feeding platform. Den is a burrow in a stream bank, but Muskrat also builds lodge of marsh plants similar to Beaver lodge though much smaller; builds feeding platform of cut plants floating on water.

Nutria *Myocastor coypus*

Also called the Swamp Beaver, this South American rodent was introduced years ago because of its valuable fur. The wetlands of the Southeast provide ideal habitat, where alligators are its only natural predators. Nutrias are now abundant and in some places even destructive. Industrious burrowers, they can quickly undermine causeways and levees, and also consume large amounts of rice and sugarcane.

Identification 26½–55″. A large, gray-brown, aquatic rodent with a long, round, nearly hairless tail and webbed hind feet. Much larger than Muskrat.

Habitat and Range Ponds, streams, and marshes in Texas, Lower Mississippi Valley, and in scattered colonies elsewhere.

Signs Hind prints are similar to Muskrat's but with webs showing between 4 long toes; fore prints smaller and rounder, with 5 pointed toes. Droppings are long and cylindrical, left on bank or feeding platform. Den is a burrow in a bank, not a lodge. Builds floating feeding platform of marsh plants, 5–6″ across. Leaves trails through marsh, and gives loud, grunting calls at dusk.

Porcupine *Erethizon dorsatum*

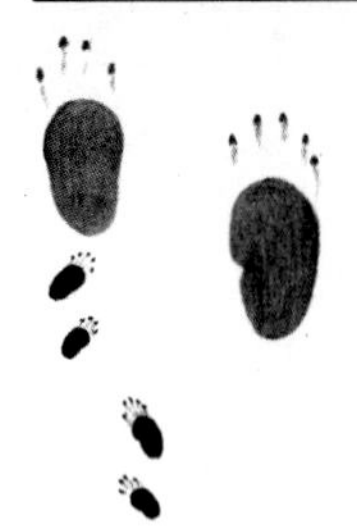

Sluggish and relying on its sharp quills for safety, a Porcupine may spend days in a single tree, quietly eating the bark, tender shoots, and buds, before moving on to another tree. The traces of these meals—pale branches with the bark removed—are the best sign that a Porcupine has been at work. With any luck, the animal itself will not be far away.

Identification 26½–40″. A stocky, tree-dwelling mammal with conspicuous quills on its back and tail. Fur is dark brown in the East, tawny in the West.

Habitat and Range Coniferous forests and mixed woodlands in Alaska, Canada, Great Lakes region, and Northeast.

Signs Hind prints are more than 3″ long, fore prints smaller, both like small human foot, with claw tips well ahead of toe marks; feet turn in, creating pigeon-toed effect. Trail often blurred in snow, showing marks of tail swinging from side to side. Droppings are variable, often rough pellets, sometimes segmented or in strings, found in piles at base of tree. Leaves twigs with bark chewed away.

Woodchuck *Marmota monax*

The familiar "groundhog" is the only marmot found in the East. It spends the winter hibernating in its burrow, but as soon as spring arrives it resumes feeding on grass and on cultivated plants if there is a garden around. When not eating, Woodchucks often sit upright, ready to give a loud whistle if alarmed and then dive into the safety of their burrow at the first sign of danger.

Identification 20–27″. A large, chunky rodent with a short bushy tail and short ears. Brown or buff-brown, lightly grizzled with white.

Habitat and Range Meadows, farmland, and open woodlands in S. Canada and C. and NE. United States.

Signs Fore prints are 2″ long, showing 4 long toes; hind prints similar but longer when heel leaves impression in soil, and showing 5 toes. Trail is usually found in disturbed earth around burrow. Droppings are variable, usually more than 1″ long. Den is a burrow, often surrounded by earth, but back entrances may be flush with the ground and hidden. Makes trails through grass to and from main entrance of burrow.

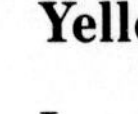
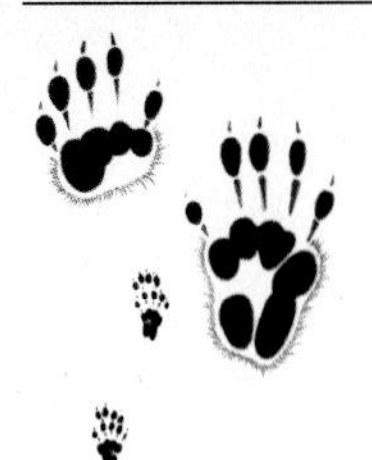

Yellow-bellied Marmot *Marmota flaviventris*

In most of the West, this colorful marmot is the mountain equivalent of the Woodchuck of eastern meadows. In its stony, alpine habitat, it leaves few signs other than its burrow. Like the Woodchuck, it feeds during the day and frequently watches for danger, choosing an exposed boulder as its lookout.

Identification 15–28″. A large, chunky rodent with a short, bushy tail and short ears. Yellowish brown with yellow belly and white patches on face; feet often dark brown.

Habitat and Range Rocky mountain slopes from SW. Canada to California and N. New Mexico.

Signs Tracks are similar to Woodchuck's, with fore prints showing 4 long toes; hind prints longer when heel leaves impression in soil, and showing 5 toes. Trail is usually found in soft earth on a ledge. Droppings are variable, usually more than 1″ long. Den is often hidden among rocks, but is sometimes a burrow 8–9″ wide, surrounded by earth on a ledge. Gives loud, far-carrying chirp or whistle when alarmed.

Hoary Marmot *Marmota caligata*

The gray color of the Hoary Marmot blends well with the rocks in its high-mountain habitat. Alert for danger as all marmots are, it is likely to be heard before it is seen. Even after it gives its alarm whistle, it may stay on its boulder, as if confident that the intruder will not be able to spot it. Only if approached will it disappear among the rocks or dive into its burrow.

Identification — 26–30″. A large, chunky rodent with a short bushy tail and short ears. Gray with blackish feet and black on crown and sides of neck; tail often reddish.

Habitat and Range — Rocky mountain slopes in Alaska, W. Canada, Washington, Idaho, and W. Montana.

Signs — Tracks are similar to Yellow-bellied Marmot's but larger. Trail is usually found in soft earth on a ledge. Droppings variable, usually more than 1″ long. Den is often hidden among rocks, but is sometimes a burrow 9–15″ wide, surrounded by earth on a ledge. Gives loud, piercing whistle when alarmed. Signs are best distinguished from Yellow-bellied Marmot's by range.

Virginia Opossum *Didelphis virginiana*

The only pouched mammal, or marsupial, in North America, the Opossum looks rather like a large rat with big ears. It is mainly nocturnal, coming out under cover of darkness to forage for fruit, insects, eggs, and carrion. It is an expert swimmer and climber, and is active throughout the year. In cold winters its ears and tail tip often become frostbitten and tattered, adding to its already unkempt appearance.

Identification 31–33″. A chunky and shaggy animal with a long, pointed snout, a hairless tail, and large, hairless ears. Fur ranges from gray grizzled with white to blackish.

Habitat and Range Woodlands, farmland, and suburban areas in E. and C. United States west to Arizona; introduced in Pacific states.

Signs Hind prints are about 2″ wide, with 4 long "fingers" and with "thumb" pointing inward at right angles to 3 front toes, and outer toe smaller and pointed outward; fore prints smaller, with 5 toes evenly spaced. Trail usually found in mud; may show dragmark of tail. Droppings are variable, depending on diet.

Brush Rabbit *Sylvilagus bachmani*

In much of western Oregon and northern California, this is the only small rabbit. Farther south, the paler Desert Cottontail can also be found, but it prefers drier, open valleys rather than hillsides covered with thick chaparral. The Brush Rabbit forages at all hours, and is usually seen feeding within a few feet of thick cover. When it bounds away, it does not show the white tail of a cottontail.

Identification 12½–14½″. A small, brown rabbit with relatively short ears and a small, inconspicuous tail.

Habitat and Range Dense thickets and chaparral-covered areas in California and coastal Oregon.

Signs Hind prints are 3″ long; fore prints smaller and rounder, and placed behind paired hind prints when running, producing a distinctive set of 4 tracks. Droppings are small, round pellets, usually in clusters. Leaves network of narrow runways between feeding and resting places. Clipped shoots are cleanly cut on a slant; deer's are ragged or crushed at tip.

Marsh Rabbit *Sylvilagus palustris*

In its wetland home, this well-named species is mainly nocturnal, unlikely to be seen unless it is startled out of its daytime hiding place. When it moves about it may hop like other small rabbits, but also it picks its way along muddy trails through the marsh by walking, leaving a trail of small, well-defined, four-toed prints quite unlike the familiar trail left by a hopping cottontail. An accomplished swimmer, the Marsh Rabbit can sometimes be found on floating masses of vegetation.

Identification 14–16″. A small, marsh-dwelling rabbit, darker and with shorter ears than an Eastern Cottontail, with dark feet and little or no white visible in the tail.

Habitat and Range Swamps and marshes in SE. United States from Virginia to Florida and Alabama.

Signs Hind prints are 3–4″ long; fore prints smaller and rounder, usually placed behind paired hind prints when running, producing a distinctive set of 4 tracks; also walks, leaving trail of small prints with 4 toes. Droppings are small, round pellets, often left on logs. Leaves network of trails in marshy habitat.

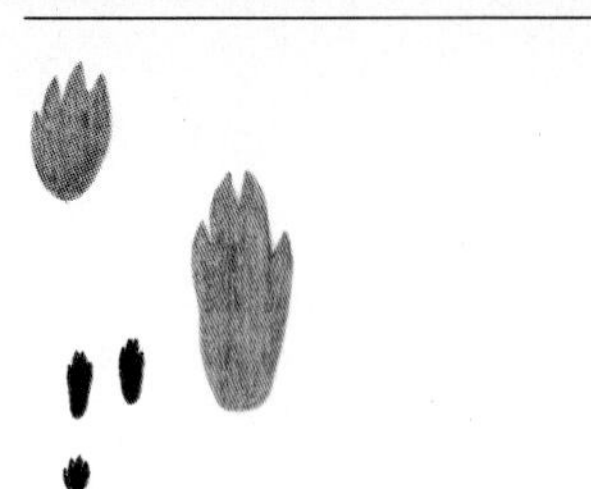

Eastern Cottontail *Sylvilagus floridanus*

In much of eastern North America, this is the familiar cottontail, easily identified by its white tail as it hops away to safety. The best time to look for these small rabbits is at dusk, when they venture out to feed on lawns, grassy roadsides, and in meadows. During the day, they often hide in brush piles.

Identification: 14–18″. A small, gray-brown rabbit with white feet and a rusty nape. Tail white below, conspicuous when the animal is running.

Habitat and Range: Woodlands, thickets, fields, and suburban areas in E. and C. United States; also in Arizona.

Signs: Hind prints are 3–4″ long; fore prints smaller and rounder, and placed behind paired hind prints when running, producing a distinctive set of 4 tracks. Droppings are small, round pellets, usually in clusters. Leaves network of narrow trails between resting places. Clipped shoots are cleanly cut on a slant; deer's are ragged or crushed at tip. Gnaws bark from the base of sumacs and other soft-barked trees, leaving tooth marks larger than those of mice.

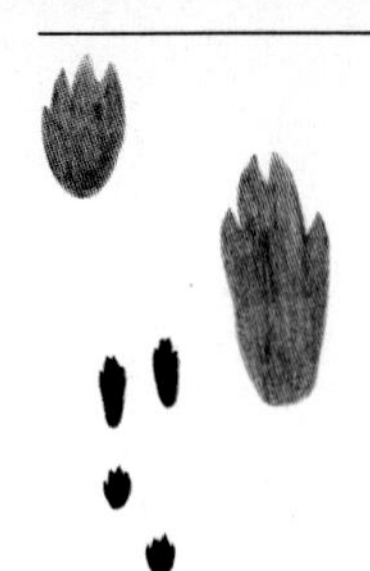

Desert Cottontail *Sylvilagus audubonii*

Throughout much of the West, this is the familiar cottontail of the dry lowlands, feeding mainly at night and leaving its tracks in soft sand. When pursued by a fox or Coyote, it darts first in one direction and then another. The trail of the predator can usually be found as well, evidence of the animal's effort to keep up with the zigzag trail left by the fleeing rabbit.

Identification 12–15″. A pale, yellow-gray rabbit with relatively long ears and a white tail, conspicuous when the animal is running.

Habitat and Range Dry open woodlands and grassy areas from California east to Great Plains.

Signs Hind prints are about 3″ long; fore prints smaller and rounder, and placed behind paired hind prints when running, producing a distinctive set of 4 tracks. Droppings are small, round pellets, usually in clusters, and often placed on log. Den is often a burrow. Clipped shoots are cleanly cut on a slant; deer's are ragged at tip. Gnaws bark from base of soft-barked trees and shrubs.

Snowshoe Hare *Lepus americanus*

This large rabbit is the favorite food of several northern predators, including the Lynx and Coyote. Its wide hind feet—the "snowshoes"—enable it to travel swiftly over the snow. The life of such a prey animal is a hard one, but that of the Lynx is even harder, because when numbers of hares decline, as they do periodically, the Lynx is faced with starvation.

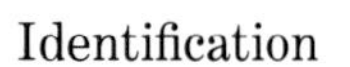

Identification 17½–21″. A forest-dwelling rabbit with large hind feet. Brown in summer, all white in winter.

Habitat and Range: Coniferous forests in Alaska, Canada, and Northeast, south in mountains to California, New Mexico, and North Carolina.

Signs: Hind prints are 4–5″ long, longer in snow and wider in front than cottontail's; fore prints smaller and rounder, and placed behind paired hind prints when running, producing a distinctive set of 4 tracks. Droppings are round pellets, usually larger than cottontail's. Leaves network of trails in snow and tufts of soft fur snagged on low branches. Clipped shoots are cleanly cut on a slant; deer's are ragged at tip.

Black-tailed Jackrabbit *Lepus californicus*

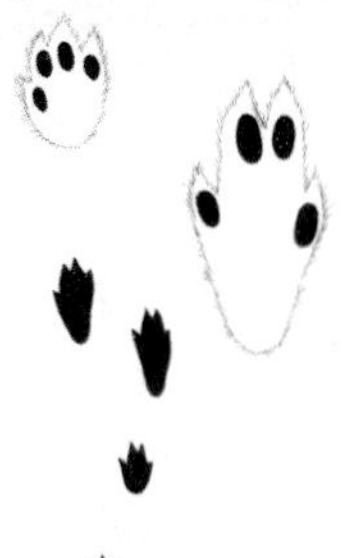

Jackrabbits are really hares; their young are not blind and helpless at birth like those of rabbits and cottontails, but have their eyes open and can soon leave the nest. Jackrabbits are built for speed. Their legs are long and slender, and when they run, their feet often hit the ground only at their tips. This produces a distinctive trail, and also enables a jackrabbit to reach a speed of 35 miles an hour.

Identification 22–24″. A large, gray-brown rabbit of open country, with very long legs and large, black-tipped ears. Tail black on top.

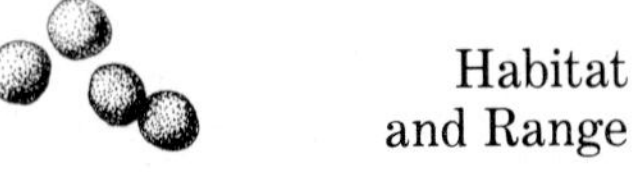

Habitat and Range Plains, meadows, and farmland in W. and SW. United States.

Signs Hind prints are 2–3″ long, with less of a "heel mark" than cottontail's; fore prints smaller and rounder, placed behind paired hind prints when running, producing a distinctive set of 4 tracks. Droppings are round pellets. Leaves network of trails between feeding and resting places. Lives in a shallow scrape (depression) under a thicket.

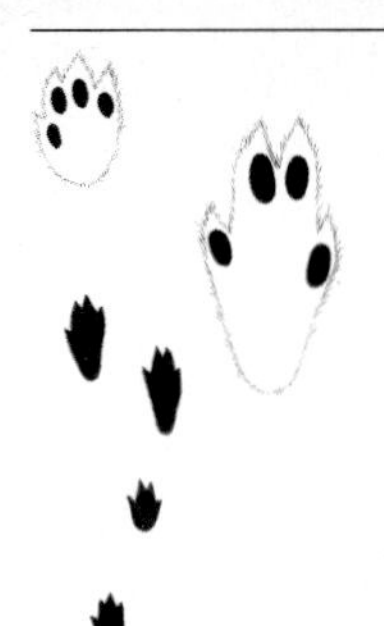

White-tailed Jackrabbit *Lepus townsendii*

A northern cousin of the Black-tailed Jackrabbit, this species lives in an area covered by snow in winter. Its white winter coat makes it harder to see, a valuable asset for an animal relentlessly pursued by foxes, Coyotes, hawks, Golden Eagles, and large owls. It is even faster on its feet than the somewhat smaller Black-tailed Jackrabbit, covering as much as 40 feet in a single bound and reaching a speed of 45 miles an hour when hard-pressed. These two jackrabbits can often be found together in the northern part of the Black-tail's range, but the White-tail ranges higher into the mountains and is more partial to truly barren country.

Identification 23–25″. A large, gray-brown rabbit of open country, with very long legs and large, black-tipped ears. Tail all white. Turns white or pale gray in winter.

Habitat and Range Grasslands and sagebrush plains from Washington, Oregon, and California east to N. Great Plains, Minnesota, Wisconsin, and Iowa.

Signs Tracks, droppings, and other signs similar to those of Black-tailed Jackrabbit.

Nine-banded Armadillo *Dasypus novemcinctus*

Protected by a covering of hard plates, armadillos forage in broad daylight as well as at night. They become so preoccupied with their search for food that they can be approached easily. An armadillo is equipped with stout claws that enable it to dig burrows quickly and to tear up the soil in search of insects.

Identification 31–34″. Unmistakable. A short-legged animal covered with an armor of horny plates and scales. Ears large; tail long, tapering, and covered with scales.

Habitat and Range Thickets and open woodlands from Texas and SE. Kansas east to Georgia and Florida.

Signs Tracks are rabbitlike, but with toes more widely spread and well defined, and with claw marks showing clearly, making them easy to identify. Hind prints show 4 toes, fore prints 5 toes. Droppings are round or oblong pellets, often containing clay. Den is a burrow, 6–8″ wide, in bank or hillside. When foraging, often leaves patches of torn-up soil or leaf litter like those of a skunk. A dug-up anthill is a good sign that an Armadillo is present.

Pika *Ochotona princeps*

Distantly related to rabbits, the Pika blends so well with its background of jumbled rocks that it is almost always heard before it is seen. Watching from a favorite perch atop a boulder, it utters a high-pitched, barking call at the slightest sign of an intruder—only then is a visitor likely to notice it. Although winters are cold in its alpine habitat, Pikas do not hibernate. Instead, they spend the late summer collecting grass and setting it out in little piles in the sun. When this "hay" has cured, the Pikas carry it down among the rocks, where it provides food during the long winter months.

Identification 7–8½″. A small, brown, stocky animal with short ears and tail, always found among rocks.

Habitat and Range Rocky slopes in high mountains of SW. Canada and United States, south to California and New Mexico.

Signs Tracks are seldom found except in snow or mud; fore prints have 5 toes, but only 4 usually show; hind prints have 4 toes, but are not elongated like a rabbit's. Droppings are black, sticky pellets. Best sign is a small pile of grass, placed on rocks to cure in the sun.

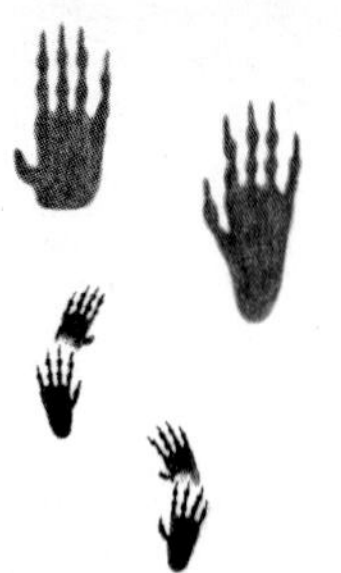

Mountain Beaver *Aplodontia rufa*

Unrelated to the Beaver, this small mammal is more like a large, forest-dwelling pocket gopher. It spends most of its time in its complex system of tunnels, leaving only to collect small green plants for food. In winter, it eats dried plants it has stored below ground, and also ventures out to forage on soft bark, the foliage of cedar, hemlock, and Douglas fir, and, especially, the fronds of evergreen ferns.

Identification 12½–17″. A small, brown, apparently tail-less rodent with short legs.

Habitat and Range Moist woodlands and swamps along coast of British Columbia, Washington, Oregon, and California; also in Sierra Nevada.

Signs Tracks are less than 2″ long, narrow with 5 long toes on both front and hind prints; inner toe on front feet may not show. Den is a burrow up to 19″ across, surrounded by a fan of disturbed earth and often covered with fern fronds. In late summer, places fern fronds and other plants on logs or rocks to dry before storing them underground. Tunnels under snow during winter; when snow melts in spring, earthen cores remain on ground.

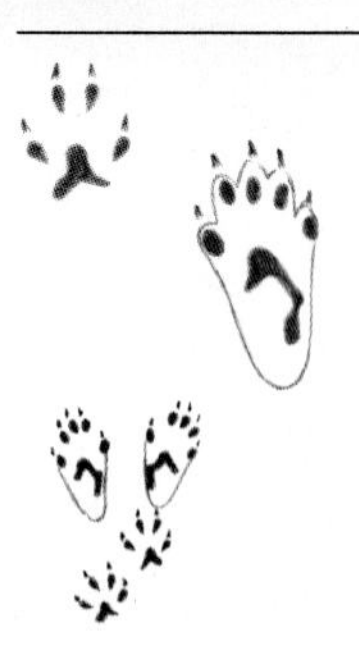

Eastern Chipmunk *Tamias striatus*

The only chipmunk in most of the East, this little member of the squirrel family is easily recognized by its stripes as it darts along stone walls, pops in and out of its burrows, and occasionally climbs trees in search of acorns and other nuts. It is highly territorial, and can be heard proclaiming ownership of a patch of woods with a loud, persistent chipping. Although chipmunks do not enter true hibernation, they spend most of the winter sleeping lightly in their tunnels and feeding on a cache of stored nuts and seeds.

Identification 9–11″. A small, ground-dwelling squirrel, with bold stripes on the back and face, and a reddish lower back.

Habitat and Range Woodlands and suburban areas in S. Canada and E. United States, west to Great Plains.

Signs Tracks are usually seen in mud; seldom found in snow; hind prints are nearly 2″ long when heel mark shows, with 5 toes; fore prints smaller and rounder, with 4 toes showing. Trail of hopping animal has hind prints in front of fore prints. Den is a burrow 2″ wide in bank or on lawn. Often leaves nutshells on rocks or logs.

Least Chipmunk *Tamias minimus*

The palest and most widely distributed of the 21 species of chipmunk in the West, this species is the only one likely to be found in dry lowlands. It climbs more readily than its larger eastern cousin, and sometimes builds nests of grass and leaves in trees. It prefers nuts, seeds, berries, and grass, but will also eat mushrooms, insects, and even an occasional mouse.

Identification 7–9″. A small, ground-dwelling squirrel, with stripes on the face and back; the stripes on the back extend to the base of the tail. Sides are reddish in the East, paler in the West. Other western chipmunks are darker.

Habitat and Range Rocky slopes, pine woodlands, and sagebrush plains in S. Canada east to Great Lakes region, and south throughout mountains of W. United States.

Signs Tracks are usually seen in fine dust; hind prints are about 1¼″ long when heel shows, with 5 toes; fore prints smaller and rounder, with 4 toes showing. Trail of hopping animal has hind prints in front of fore prints. Den is a burrow 2″ wide in bank or a small nest of leaves and grass in a shrub or small tree. Often leaves nutshells on rocks or logs.

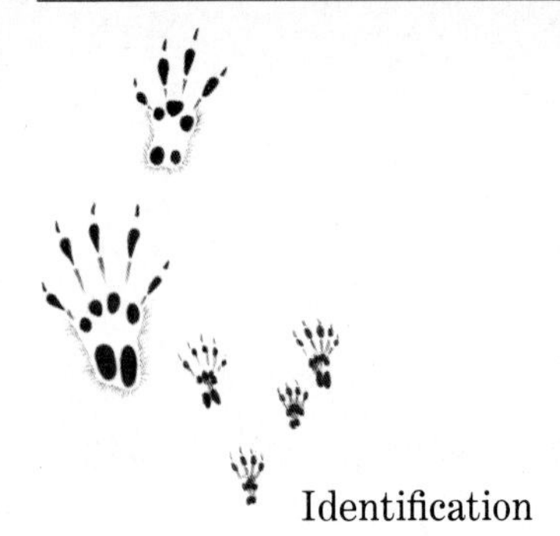

Thirteen-lined Ground Squirrel *Spermophilus tridecemlineatus*

This common ground squirrel is most often seen standing up at the entrance of its burrow, a habit that led to the name "picket pin" after the stakes used to anchor tents or tether horses. Its burrows can be skillfully concealed, but picket pins rarely move far from home. So if you see one of these squirrels eyeing you from a patch of grass, you can be sure it has a burrow close by.

Identification 8–14″. A small, sandy-brown, ground-dwelling squirrel with darker brown stripes on the back; stripes contain rows of pale spots.

Habitat and Range Plains, prairies, and lawns in Great Plains and Great Lakes region, south to Texas.

Signs Tracks are usually seen in dust; hind prints are about 1½″ long when heel shows, with 5 toes; fore prints smaller and rounder, with 4 toes showing. Droppings are variable, usually elongated, ½″ pellets. Den is a burrow with a 2″ entrance, surrounded by disturbed earth and with runways radiating away from it, but burrows are often well concealed in grass and hard to find.

California Ground Squirrel *Spermophilus beecheyi*

Like most other ground squirrels, this species lives in colonies that may last many generations. Ranging from high mountain meadows to grassy places within sight of the Pacific Ocean, the California Ground Squirrel usually locates its colonies on gentle, well-drained slopes. Most adults hibernate between November and February, and also withdraw into their burrows during the hottest summer months. At other times, they spend most of their existence gathering food and fighting.

Identification — 17–19″. A large, ground-dwelling squirrel, brown above, paler below, with vague whitish flecks on the back and a grayish patch on the shoulders and sides of the neck.

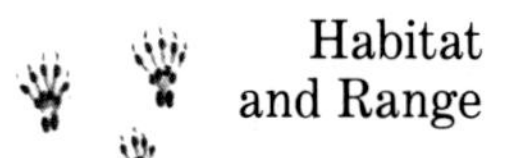

Habitat and Range — Fields, rocky slopes, and barren areas in S. Washington and W. Oregon, and W. California.

Signs — Tracks are usually seen in dust; hind prints are about 1¾″ long when heel shows, with 5 toes; fore prints smaller and rounder, with 4 toes showing. Droppings are variable, usually elongated, ½″ pellets. Den is a burrow 2″ wide, with an entrance mound; several such burrows are usually found together.

Golden-mantled Ground Squirrel *Spermophilus lateralis*

Sometimes called the "copperhead," this brightly colored ground squirrel avoids the open, grassy country preferred by other ground squirrels. It does not live in colonies, and, sleeping only lightly during the winter, stores food and may occasionally tunnel up through the snow. It is especially common in parks, around campsites, and near groups of summer houses in the woods, where it can supplement its natural diet of seeds, fruit, insects, and carrion with table scraps and handouts.

Identification — 10–12″. A medium-sized, ground-dwelling squirrel, larger than a chipmunk and with stripes only on the back and with a reddish tinge (the "mantle") on the face and neck.

Habitat and Range — Coniferous forests and brushy areas in mountains of W. Canada and W. United States.

Signs — Hind prints are about 1¼″ long when heel mark shows, with 5 toes; fore prints smaller and rounder, with 4 toes showing. Droppings are variable, usually elongated, ½″ pellets. Den is a burrow 3″ wide, usually concealed under a log, stump, or boulder.

Black-tailed Prairie Dog *Cynomys ludovicianus*

Among the most social of mammals, prairie dogs—so named because of their barking call—always live in "towns." Large towns are divided into jointly defended "wards," which in turn are made up of family groups, or "coteries." The sharp bark serves as a warning that sends every prairie dog diving into its burrow.

Identification 14½–16½″. A stocky, yellow-brown, ground-dwelling squirrel with a short, slender, black-tipped tail. White-tailed Prairie Dog (*C. leucurus*) of mountain valleys is similar, but has a white tail tip and dark smudges above and below the eyes.

Habitat and Range Grasslands and prairies in Great Plains, from S. Canada south to W. Texas.

Signs Hind prints are about 1¼″ long with 5 toes; fore prints smaller, with 4 toes showing. Droppings are variable, usually elongated, ½″ pellets; may be a connected string of small pellets. Den is a conical burrow 1′ or more high and about 2′ wide; such burrows are usually found in groups that may occupy 100 acres or more.

Gray Squirrel *Sciurus carolinensis*

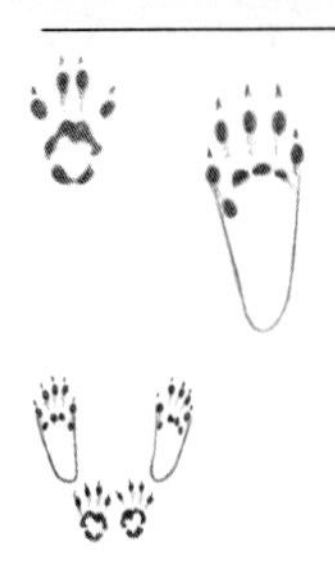

This adaptable squirrel is one of our most abundant mammals, thriving in city parks as well as in remote forests. City squirrels are bold in their quest for bread and other handouts, but their country cousins, accustomed to being hunted and living among a variety of predators, can be shy and hard to find.

Identification 17–20″. A gray or buff-gray tree squirrel with whitish underparts and a bushy tail bordered with white-tipped hairs. Black individuals occur in some areas.

Habitat and Range Forests, woodlands, and suburban areas in S. Canada and E. United States, west to Great Plains.

Signs Hind prints are about 2¼″ long, with 5 toes showing; fore prints rounder and smaller, about 1″ long. Trail of hopping or running animal has hind prints in front of fore prints; such a trail often extends between trees. Den is placed in a hollow tree or in a loose, ball-shaped nest of leaves among branches; summer nests are flat saucers of leaves. Often disturbs leaf litter when searching for food. Chews nuts open, leaving large, ragged holes in the shells, and often bites away the large end of acorns.

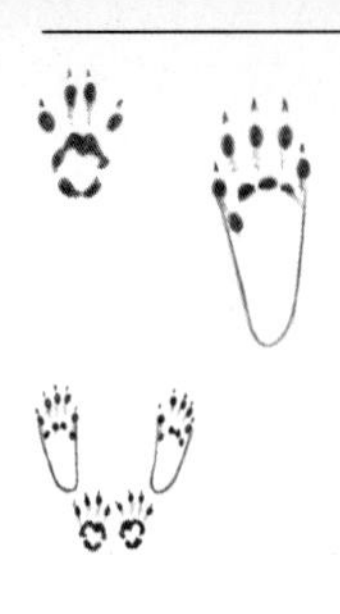

Western Gray Squirrel *Sciurus griseus*

While the Gray Squirrel of the East prefers forests of oaks and hickories, this species lives mainly among conifers or in forests containing both pines and oaks. In the fall, when acorns appear in the oak forests that grow at lower elevations than the conifers, these squirrels may travel down the slopes to harvest them.

Identification 18–23″. A gray tree squirrel with white underparts and a bushy tail that is white below and a mixture of gray, white, and black hairs above.

Habitat and Range Forests and woodlands from W. Washington and Oregon south to S. California.

Signs Hind prints are about 2¼″ long, with 5 toes showing; fore prints rounder and smaller, about 1″ long. Trail of hopping or running animal has hind prints in front of fore prints; in snow, such a trail often extends between trees. Den is placed in a hollow tree or in a loose, ball-shaped nest of leaves among branches; summer nests are flat saucers of twigs and shredded bark. Often strips bark from branches of pines. Chews nuts open, leaving large, ragged holes in the shells.

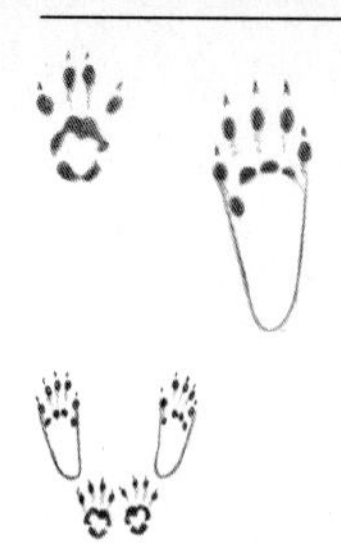

Fox Squirrel *Sciurus niger*

Like the Gray Squirrel, the Fox Squirrel is adaptable, tame, and common in parks and suburbs. The two species are similar in diet and habits, but the smaller Gray Squirrel is more aggressive and successful and usually wins out in competition for food and nesting places.

Identification 18–27″. A large tree squirrel, usually gray or buff-gray above and yellow or pale orange below. Southern animals grizzled, with black and white marks on face. Mid-Atlantic animals even grayer than Gray Squirrel.

Habitat and Range Forests, woodlands, and suburban areas in E. United States, west to Dakotas, Colorado, and Texas; absent from Northeast.

Signs Hind prints are about 2½″ long, with 5 toes; fore prints rounder and smaller, about 1″ long. Trail of hopping animal has hind prints in front of fore prints. Den is placed in a hollow tree or in a loose, ball-shaped nest of leaves among branches. Often carries nuts and corncobs to a favorite eating place, leaving a pile of litter below. Chews nuts open, leaving large, ragged holes in the shells, and often bites away the large end of acorns.

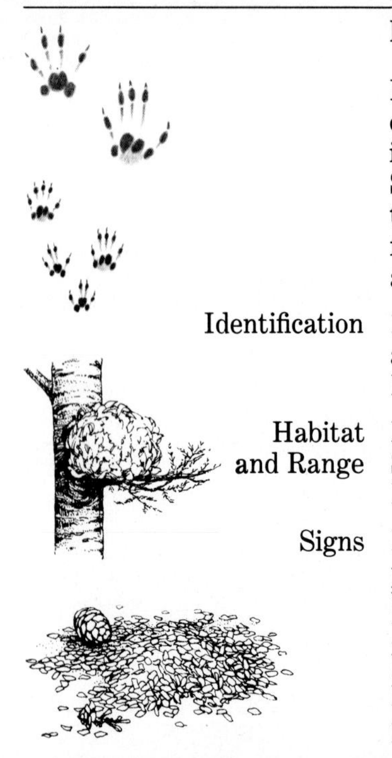

Red Squirrel *Tamiasciurus hudsonicus*

Nervous and agile, the Red Squirrel is found in forests of all kinds but is most abundant among conifers, where it is relatively free of competition with the larger Gray Squirrel. Here it cannot be overlooked, because it greets the arrival of any visitor with a loud, sputtering chatter. It forages both in the trees and on the ground, often burying a supply of pinecones for use during the winter.

Identification — 12–14″. A small tree squirrel, rust-red or yellowish above and whitish below. Douglas' Squirrel (*T. douglasii*) of the Northwest similar but grayish or orange below.

Habitat and Range — Forests and woodlands in Alaska, Canada, and N. United States, south in mountains to Arizona, New Mexico, and North Carolina.

Signs — Hind prints are about 1½″ long, with 5 toes showing; fore prints are about ¾″ long, with 4 toes showing. Trail of bounding animal has hind prints in front of fore prints. Makes nest of grass and shredded bark, usually in conifer, but sometimes lives in burrow at base of stump or under log. Brings cones and nuts to favorite feeding place, leaving pile of chaff, cone scales, and litter below.

Southern Flying Squirrel *Glaucomys volans*

Feeding at night and spending the day asleep in a tree cavity, these skillful gliders are often much more common than occasional sightings would suggest. They are especially fond of hickory nuts; in the fall, neatly opened shells can often be found on the ground under the trees, a sure sign that flying squirrels fed there during the night.

Identification — 9–9½″. A small tree squirrel, gray-brown above and white below, with a loose fold of skin between front and hind legs. Northern Flying Squirrel (*G. sabrinus*) of Canada and the northern states larger and browner.

Habitat and Range — Forests and woodlands in E. United States, except in N. New England and Florida, and west to Great Plains.

Signs — Hind prints are about 1½″ long, with 5 toes showing; fore prints are about ¾″ long, with 4 toes showing. Trail often blurred by fold of skin between front and hind legs. Animal landing on snow leaves distinctive print showing all 4 feet and folds of skin. Usually lives in tree cavities, but may make summer nest of leaves, smaller than Gray Squirrel's. Chews nuts more neatly than other squirrels, often leaving single small opening in shell.

Shrews Family Soricidae

Tiny and mouselike, shrews have such a high rate of metabolism that they will starve to death in a matter of hours. Day and night, they search feverishly through leaf litter and in tunnels for insects, earthworms, snails, slugs, and even mice. A rustling in the leaves on the forest floor is often the only indication of their presence, but patience will sometimes be rewarded by the sight of the shrew's sharply pointed snout as the animal peers out briefly before resuming its endless hunt for food.

Identification 3–5″. Small, mouse-sized animals with beadlike eyes, small ears partly concealed in fur, and pointed snouts. Short-tailed Shrew shown here. 31 species in North America.

Habitat and Range Woodlands, fields, marshes, and swamps throughout Alaska, Canada, and United States.

Signs Tracks are small, blurry marks in dust or snow, often showing drag mark of tail. Sometimes burrows in snow, leaving small raised ridge on surface. Den is a small burrow in soil, with entrance the size of a dime or smaller, but often appropriates burrows of moles. May leave pile of empty snail shells under log.

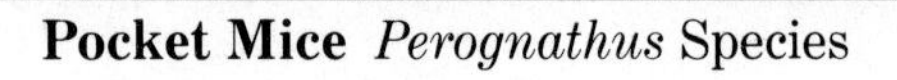

Pocket Mice *Perognathus* Species

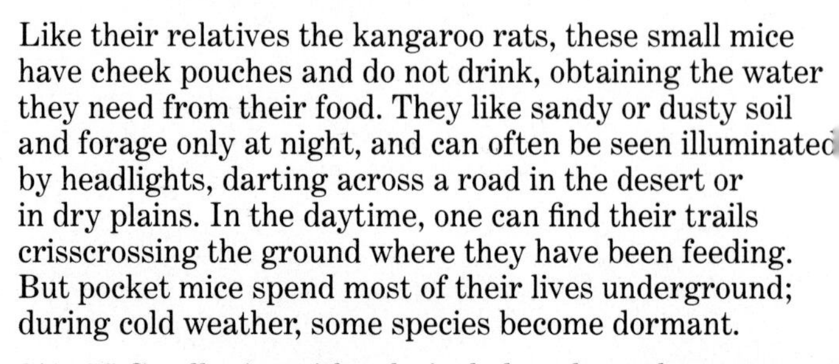

Like their relatives the kangaroo rats, these small mice have cheek pouches and do not drink, obtaining the water they need from their food. They like sandy or dusty soil and forage only at night, and can often be seen illuminated by headlights, darting across a road in the desert or in dry plains. In the daytime, one can find their trails crisscrossing the ground where they have been feeding. But pocket mice spend most of their lives underground; during cold weather, some species become dormant.

Identification 3½–5″. Small mice with relatively long legs, short ears, and fur-lined cheek pouches. The color varies from dark brown to pale pinkish buff above, and from buff to white below. Shown: *P. penicillatus.* 18 species in North America.

Habitat and Range Grasslands, deserts, and dry sandy areas throughout SW. United States.

Signs Tracks are tiny round prints in fine dust, usually poorly defined. A set of 4 tracks is about 1″ wide. Lives in ¾–1″ burrow at base of shrub or in bank, usually with mound of fine soil or sand at entrance and sometimes with trails leading away.

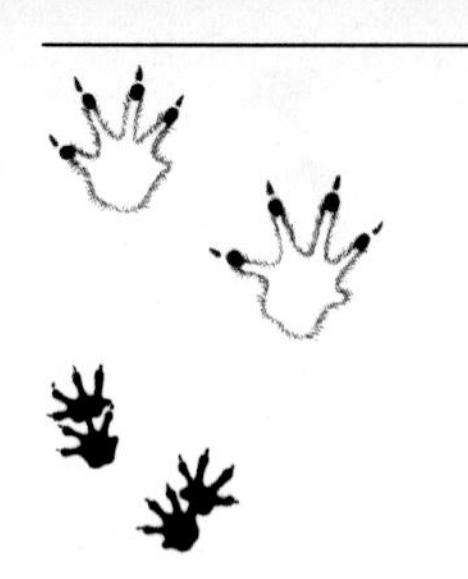

Grasshopper Mice *Onychomys* Species

Named for the grasshoppers they eat, these mice are voracious little predators, taking not only grasshoppers but beetles, other insects, and even small rodents. One species is called the Scorpion Mouse because scorpions form an important part of its diet. Perhaps because they are predators, they are not as numerous as other, more strictly vegetarian mice. Adults are highly aggressive toward other grasshopper mice, but it is believed that a male and female share the same territory all year, living together in a burrow or in any sheltered place at ground level.

Identification 5–7½″. Stocky, gray or pinkish-buff mice with short, white-tipped tails. The shorter, white-tipped tail distinguishes them from Deer Mouse and its relatives. Shown: Northern Grasshopper Mouse. There are 3 very similar species.

Habitat and Range Grasslands and sagebrush plains throughout lowland W. United States except along Pacific coast and in interior Northwest.

Signs Tracks are about ½″ wide, shorter and more rounded than those of the Deer Mouse and its relatives, usually found in mud or fine dust; fore prints and hind prints may overlap.

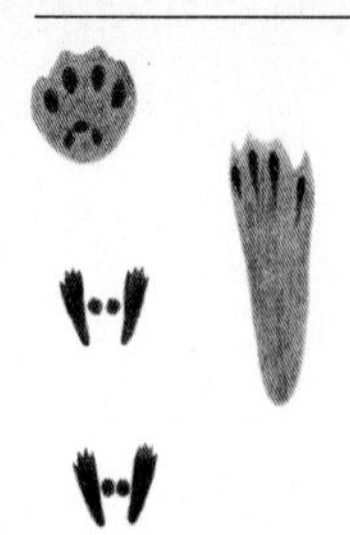

Ord's Kangaroo Rat *Dipodomys ordii*

In almost any desert area, kangaroo rats are common and conspicuous nocturnal foragers, easily identified by their habit of hopping on their long hind legs. They feed mainly on seeds, which they carry in large, external pouches in their cheeks, and can get water from the food they eat, rarely or never drinking.

Identification 9–11″. A small jumping rodent with long hind legs, short front legs, and a long tail more furry at the tip than at the base; buff-brown above and white below, with black and white stripes along the tail. The 14 other species in North America are darker or have paler tail stripes.

Habitat and Range Deserts and dry sandy areas in S. Canada and most of W. United States.

Signs Tracks of slow-moving animal show hind prints about 1½″ long, with tiny fore prints placed between them and often with drag mark left by tail. Bounding animal leaves smaller, paired hind prints, without heel marks and usually without fore prints or tail mark. Trails are often abundant in desert sand. Droppings are small, oblong, and brown or green. Den is a system of burrows in a mound of sand.

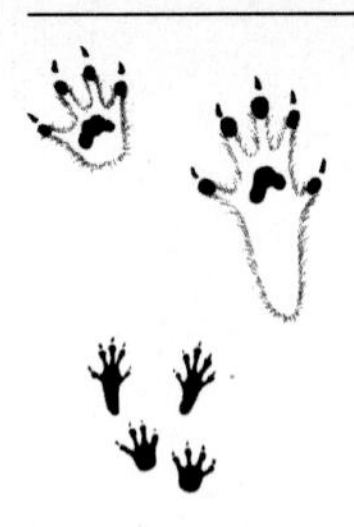

Deer Mouse *Peromyscus maniculatus*

The Deer Mouse is one of more than a dozen *Peromyscus* species, one or another of which can be found in almost any habitat in North America. All of them have white feet, white underparts, and long, slender tails. They forage on the ground, but most can climb, and some nest in bushes and trees. Several species, including the Deer Mouse, enter buildings during the winter.

Identification 6–9″. A long-tailed, slender mouse, grayish or reddish above and white below and on the feet. Tail longer than the body and usually dark above and pale below. White-footed Mouse prefers woods and has shorter tail.

Habitat and Range Woodlands, prairies, and fields in Canada and United States except in S. New England and Southeast.

Signs Hind prints are ⅝″ long, with 5 toes showing; fore prints smaller, with 4 toes; both feet have outer toes pointing to the side. Hind prints are placed in front of fore prints. Often lives in abandoned bird's nest, adding a roof over it with new nesting material, or in small burrow in ground. Leaves piles of nuts, cherry pits, or seeds under logs or in nooks in buildings.

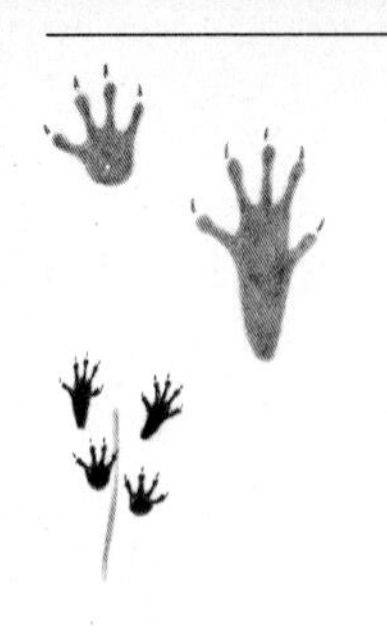

White-footed Mouse *Peromyscus leucopus*

Often called the "Wood Mouse" to distinguish it from the House Mouse, this species is more partial to woodlands than its cousin the Deer Mouse. Females are territorial during the breeding season, which occurs in spring and again in fall in the North, but may last all year in the South. Like the Deer Mouse, this species is active in winter and is usually the most abundant mammal wherever it occurs; populations of a dozen per acre are not unusual.

Identification 6–8″. A long-tailed, slender mouse, reddish above and white below and on the feet. Tail shorter than body and usually dark both above and below. Deer Mouse prefers more open country and has longer tail.

Habitat and Range Woodlands and thickets in S. Canada and E. United States west to Rocky Mountains, except in Southeast.

Signs Signs similar to Deer Mouse's. Hind prints are ⅝″ long, with 5 toes showing; fore prints smaller, with 4 toes. Hind prints are placed in front of fore prints. Often lives in abandoned bird's nest, building a roof over it with new nesting material, or in cavity in tree. Leaves caches of nuts cherry pits, or seeds under logs or in nooks in buildings.

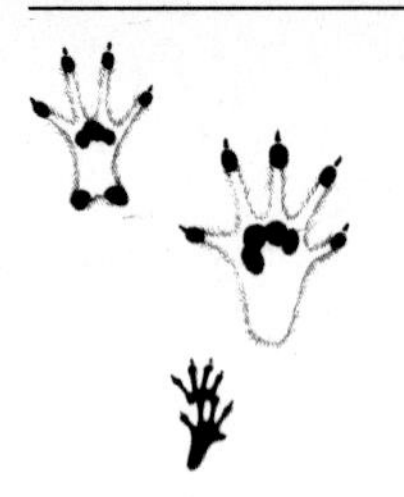

Hispid Cotton Rat *Sigmodon hispidus*

Cotton rats are easily identified by their coarse hair and short ears. They feed both day and night, taking a wide variety of plant foods as well as insects, birds' eggs, crayfish, and even fiddler crabs. They also do considerable damage to crops such as sugarcane and sweet potatoes. Breeding all year, they are abundant wherever they occur.

Identification 8–14″. A short-eared rat with coarse hair; gray-brown mixed with buff and black above, whitish below. Three related species, also coarse-haired and short-eared, occur in the Southwest. Tail shorter than in other rats.

Habitat and Range Grasslands and weedy areas in S. United States west to Arizona.

Signs Tracks are usually found in mud; similar to those of Eastern Woodrat. Hind prints are about 1″ long, with 5 long, well-separated toes; fore prints shorter, with 4 long toes. Droppings are dark, oblong, about ½″ long. Nest is a small mass of grass in a cuplike depression in the ground. Often makes runways on the surface or under grass in meadows, larger than those of voles.

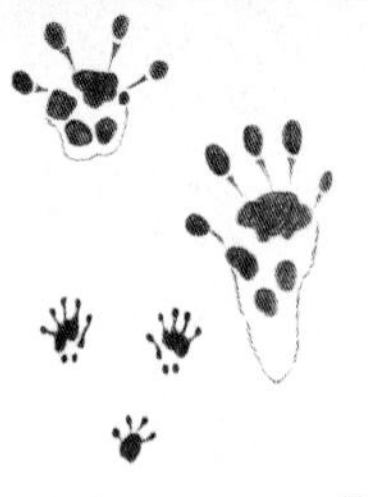

Eastern Woodrat *Neotoma floridana*

Woodrats are attracted to shiny objects, and as they gather material for their large nests, they often pick up such things and leave behind the stick they were carrying. This habit has given them the names "trade rat" and "pack rat," and has led to many tales, including one about a woodrat that stole a half dollar but brought back two quarters.

Identification 13–16″. A medium-sized, gray-brown rat with whitish underparts and feet, large ears, and a long, hairy tail. Most other woodrats grayer or have dark feet; found in the West or Southwest.

Habitat and Range Woodlands, thickets, and rocky areas from New York and South Dakota south to N. Florida and Texas.

Signs Tracks are usually found in mud; similar to those of Hispid Cotton Rat. Hind prints are about 1″ long, with 5 long toes; fore prints shorter, with 4 long toes. Makes large nest of sticks, bones, and rubbish, located on ground at base of hollow tree or hollow log, above ground anchored to limbs of trees, or in abandoned buildings. These nests have several entrances and often contain shiny objects such as key chains, cutlery, or even coins.

Southern Red-backed Vole *Clethrionomys gapperi*

Because the runways they use are on the surface of the ground, red-backed voles are seen more often than species like the Meadow Vole, whose runways are hidden under the grass. During the summer, red-backed voles prey on insects and their larvae and eat green plants and berries. In winter, they switch to a diet of durable items such as tubers, bulbs, nuts, and pits, which they often store in runways or nests.

Identification 5–6″. A small vole with soft, brown fur, small ears, and a short tail. Usually a reddish tinge in the middle of the back. Western Red-backed Vole (*C. occidentalis*) is similar, but restricted to the Northwest Coast.

Habitat and Range Wooded swamps and bogs in Canada and N. United States, south in mountains to Arizona and North Carolina.

Signs Tracks are about ½″ long and seldom noticed; hind prints have 5 toes, fore prints have 4 toes. Runway systems are not as extensive as those of the Meadow Vole, but are easier to find on the forest floor as they wind through leaf litter or beds of moss and under logs or rocks; these passageways often contain bits of cut green plants.

Meadow Vole *Microtus pennsylvanicus*

This is the abundant "field mouse," which feeds both day and night in its elaborate network of narrow passageways that run under the grass and through the snow. A mainstay of the diet of many predators, it is hunted by hawks, owls, skunks, and foxes. Its numbers remain high, however, because females can breed when only 25 days old and produce litters throughout the year.

Identification: 5½–8″. A small vole with long fur, short ears, and a short tail; dark brown, gray-brown, or yellowish grizzled with blackish. Other voles have different ranges or are larger, yellowish on snout, or even shorter tails.

Habitat and Range: Meadows, fields, marshes, and grassy woodlands from the Arctic south to New Mexico, Missouri, and Georgia.

Signs: Tracks are about ½″ long, usually in snow; hind prints have 5 toes, fore prints have 4 toes. Jumping animal may leave prints more than 4″ apart. Makes extensive network of runways under grass; runways contain piles of grass cuttings. In winter, pushes debris into tunnels under snow; in spring, these cores, about 1″ wide, are exposed by melting snow.

House Mouse *Mus musculus*

Although it seems to have originated in Asia, where it still has wild relatives, the House Mouse became associated with humans thousands of years ago and is now found throughout the world. In agricultural areas, House Mice may live in fields and feed on crops, but they come inside when the weather turns cold in the fall and leave signs that are all too familiar. Indoors, the House Mouse can be amazingly abundant; densities as high as ten to the square yard have been reported.

Identification 6–6½″. A gray or gray-brown mouse with a long, uniformly dark, hairless tail and large ears. Usually found indoors. Deer Mice and White-footed Mice white below.

Habitat and Range Buildings and cultivated areas in S. Canada and throughout United States.

Signs Tracks are seldom noticed; similar to those of Deer Mouse. Most obvious sign is small droppings, like black grains of rice, found singly or in small groups. Makes nests in sheltered places in buildings, usually of shredded paper, cardboard, or other material. In old buildings, may make round holes in walls or near floorboards.

Norway Rat *Rattus norvegicus*

Like the House Mouse, the Norway Rat, or Brown Rat, probably originated in Asia but is now found everywhere in the world. Aggressive and destructive, it eats or contaminates grain and other stored food, and carries diseases such as tularemia, spotted fever, and bubonic plague. This species arrived in the United States at about the time of the American Revolution, and has since become much more numerous than the Black Rat. Predators have little impact on its numbers, and its populations seem to be limited mainly by the amount of available food.

Identification — 15–16″. A large rat with large ears and a scaly, hairless tail that is shorter than the head and body; gray-brown above, paler below. Black Rat has longer tail.

Habitat and Range — Buildings and cultivated areas in S. Canada and throughout United States.

Signs — Hind prints are about 2″ long, with 5 long toes; fore prints less than 1″ long, with 4 toes. Droppings are dark, ½–¾″ long. Indoors, makes nests of chewed paper, cardboard, and rags; chews round holes in walls, and leaves oily smudges near holes and along regular pathways in buildings.

Black Rat *Rattus rattus*

Also known as the Ship Rat, this species arrived in Europe and then North America long before the Norway Rat, and was the carrier of bubonic plague during the famous "Black Death" of the 14th century, which killed a third of the population of Europe. As soon as the more competitive Norway Rat arrived in North America, the Black Rat began to disappear, and today it is much less common than the Norway Rat. It is a better climber, and where the two species are found together, the Black Rat generally lives on the upper floors of buildings, with the lower floors and the cellar, as well as sewers and garbage dumps, taken over by the Norway Rat.

Identification — 15–18″. A large rat with large ears and a scaly, hairless tail that is longer than the head and body; black or brown above, paler below. Norway Rat has shorter tail.

Habitat and Range — Buildings (mainly on upper floors) in coastal cities and some inland towns in the United States.

Signs — Tracks, droppings, and other signs similar to those of Norway Rat.

Meadow Jumping Mouse *Zapus hudsonius*

Living in tall grass and traveling about by bounding, jumping mice leave few tracks. They eat many insects, but a large part of their diet consists of the seeds of grasses, which they harvest by reaching up and cutting a stem, a little at a time, until the seed head is close to the ground. These clusters of cut grass stalks are the best sign that jumping mice are around. Unlike many other mice, they go into deep hibernation during the winter, so that their tracks are not found in snow.

Identification 8–10″. A long-legged, long-tailed mouse, brown above with a yellow tinge on the sides and whitish underparts. Western Jumping Mouse (*Z. princeps*) larger and found mainly south and west of Meadow Jumping Mouse.

Habitat and Range Wet meadows, grassy marshes, and weedy areas in S. Alaska and Canada south to Colorado, Oklahoma, Alabama, and Carolinas.

Signs Tracks are rarely found. The best sign is a small pile of cut grass stems, up to 5″ long, lying among growing plants in a meadow. Makes small, globular nest of grass on ground.

Pocket Gophers *Geomys* and *Thomomys* Species

These burrowing rodents, called pocket gophers because of their large cheek pouches, spend their lives tunneling in the ground, attacking plants from below and pulling the edible parts into their burrows. They dig with their stoutly clawed front feet and with their large incisors, pushing excess soil up to the surface in a string of mounds that marks the location and direction of their burrows. They feed during both summer and winter.

Identification: 7½–14″. Stocky, burrowing rodents with large, yellow front teeth, small eyes and ears, and long, stout claws on the front feet. Shown: Plains Pocket Gopher. There are 17 species of pocket gophers in North America.

Habitat and Range: Sandy plains, pastures, and open mountain valleys in W. and S. United States.

Signs: Hind prints are about 1″ long, with 5 toes; fore prints smaller, with 4 toes and with claw marks well defined. Best signs are large, fan-shaped mounds of excavated earth, often with entrance plug, and tunnel cores, which are long masses of earth about 2″ wide, placed in burrows under snow during winter and exposed by melting snow in spring

Moles Family Talpidae

Moles are even more modified for burrowing than gophers; they have streamlined bodies, stout claws for digging, tiny, almost useless eyes, ears so small that they do not catch in a tunnel, and velvety fur that does not stick to the soil, as well as a sensitive nose equipped with whiskers for feeling their underground prey. Their tunnels near the surface create ridges of earth known as molehills.

Identification 5½–9″. Small, stout-bodied, grayish or blackish animals with short, silky fur and large front feet with stout claws, and with eyes and ears that are scarcely visible. Shown: Star-nosed Mole. There are 7 species in North America.

Habitat and Range Woodlands, fields, and suburban areas in S. Canada, E. and C. United States, and in Pacific states.

Signs Tracks and droppings seldom noticed. Best signs are molehills, long ridges heaved up by animal as it burrows near the surface of the ground, on lawns, in meadows, or in woodlands. These ridges differ from the tunnel cores of pocket gophers by being covered with leaf litter or small growing plants. Mounds of soil are often pushed out of burrows; such mounds do not contain an entrance.

Hoofprints
pages 18–36

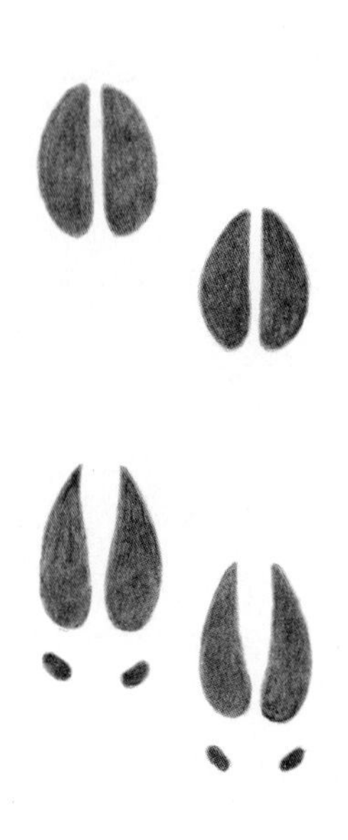

Tracks with Five Toes and Pad
pages 38–54

Tracks with Four Toes and Pad
pages 56–78

Footlike or Handlike Tracks
pages 80–102

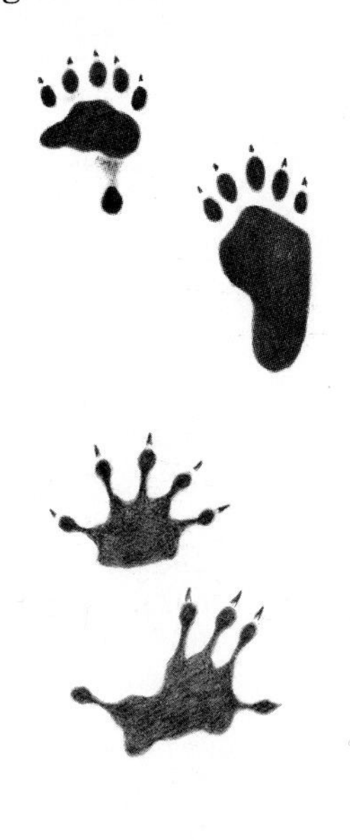

Rabbitlike Tracks
pages 104–118

Small Tracks
pages 120–176

Guide to Mammal Orders and Families

Mammal species are classified in groups called families, and related families are placed together in groups called orders. Knowing the general features of these categories can be helpful in identifying tracks and other signs.

Opossums

These belong to the order Marsupialia, the pouched animals that include the kangaroos. The Virginia Opossum, which belongs to the family Didelphidae, is our only member of this group. Marsupials are primitive animals whose young develop in a pouch. They have five long toes on each foot.

Shrews and Moles

The order Insectivora, the "insect-eaters," includes both the shrews (family Soricidae) and the moles (family Talpidae). Young shrews and moles develop inside the mother's body, like those of all other North American mammals except the Virginia Opossum, but these insect-eaters are also primitive animals. They spend most of their time out of sight—under leaves on the forest floor in the case of shrews, and burrowing in the ground in the case of moles.

Armadillos

Along with sloths and anteaters, armadillos make up the order Xenarthra, sharing features not visible on the outside of the body. Armadillos (family Dasypodidae) are best known for their armor-plated skin and their stout toes, which are equipped with strong claws for digging.

Lagomorphs

The order Lagomorpha, which means "rabbit-shaped," includes the rabbits and hares (family Leporidae) and their close kin the pikas (family Ochotonidae). Like the rodents, lagomorphs have large incisor teeth well suited for gnawing, but while rodents have a single pair of incisors on each jaw, lagomorphs have a second pair of upper incisors, tiny teeth located just behind the main pair. Rabbits and hares are built for speed, with large hind legs and powerful feet. Pikas, which live among rocks and seldom leave tracks, have smaller, more delicate feet.

Rodents

The huge order Rodentia includes thousands of species and nearly half of all North American mammals. They owe their success to their adaptability and their chisel-like incisor teeth, which grow in as fast as they are worn down. The rodents are divided into three groups based on how they use their teeth and the muscles that operate their jaws.

Squirrel-like Rodents

This important group includes not only the squirrels, chipmunks, marmots, and prairie dogs (family Sciuridae), but the peculiar Mountain Beaver (family Aplodontidae), the pocket gophers (family Geomyidae), the pocket mice and kangaroo rats (family Heteromyidae)—these latter two families are found only in North and Central America—and the Beaver (family Castoridae). The most varied of the three groups of rodents in North America, they include tree-dwelling, aquatic, surface-dwelling, and burrowing animals that outwardly have little in common.

Rats and Mice The mouselike rodents are the largest, most prolific, and least varied of the three groups. In North America they include the typical rats and mice, the voles, and the Muskrat, which is really a giant vole; all these belong to the family Cricetidae. Other members of this group are the introduced Old World Black Rat, Norway Rat, and House Mouse (family Muridae), and the long-legged, long-tailed jumping mice (family Zapodidae). All have slender and often hairless tails and relatively large ears, and all look unmistakably like rats or mice.

Porcupine and Nutria Both the native Porcupine (family Erethizontidae) and the introduced Nutria (family Myocastoridae) belong to a group that is centered in South America, where they are even more varied than the squirrel-like rodents. The best signs of the tree-dwelling Porcupine are gnawed branches, and tracks that look like small human feet.

The Nutria is aquatic, and leaves tracks that show webs between the toes.

Carnivores

The flesh-eating mammals of the order Carnivora all share long, sharp canine teeth for grasping and subduing their prey. All are strong, and are built either for speed or for stealth.

Foxes and Wolves

Relatives of dogs (family Canidae), the foxes and wolves all have long, slender legs, large ears, pointed muzzles, and tracks like those of the domestic dog. Unlike many carnivores, the dogs have long-lasting social bonds, either between members of a pair or in an extended family or "pack"; members cooperate in bringing down large prey.

Bears

North America's three bears (family Ursidae) are large, relatively ponderous carnivores that vary their meat diet with berries and roots when they can find them. Their feet are large with stout claws, and their tracks are large and broad, somewhat resembling those of a human foot.

Raccoon and Ringtail

Both the Raccoon and its relative the Ringtail (family Procyonidae) have pointed muzzles and banded tails; like the bears, they vary their diet with berries and vegetables. The Raccoon is famous for its intelligence and manual dexterity; its fingers and toes are long and well suited for holding and manipulating food, and leave tracks that are easy to recognize. The Ringtail has smaller feet with shorter toes, and leaves tracks like those of a house cat.

Weasels The large and diverse family Mustelidae includes both the smallest and the most ferocious carnivores. All are equipped with very sharp canine teeth. The smaller weasels, and the larger Mink, Fisher, and Marten, are fast-moving predators that relentlessly pursue their prey through crevices and tunnels or among the branches of trees. The long, sleek River Otter and Sea Otter have taken to the water, where they use their webbed feet for swimming. The Badger and Wolverine are stocky, broad-shouldered animals with stout claws; the Badger uses its claws for digging up ground squirrels and other burrowing rodents, while the Wolverine uses its claws, and its formidable array of teeth, to tackle prey much larger than itself. Because of their defensive, foul-smelling spray, the skunks lead more peaceful lives, puttering about in search of insects or fruit; skunks cannot move as fast as other members of the weasel family, and their tracks are more flat-footed, resembling the prints of bears in miniature.

Cats Our three wild cats (family Felidae), the Mountain Lion, Lynx, and Bobcat, all live by stealth, sneaking up on their prey or pouncing on it from concealment. They have sharp canines and even sharper claws that can be retracted. They are solitary for most of the year, and their tracks look like large versions of the house cat's.

Hoofed Mammals North America has four families of hoofed mammals (order Artiodactyla), all of them vegetarians that have long legs, designed for speed, and hooves on two toes that leave paired or "cloven" hoof marks. All tend to travel in herds or small bands.

Peccaries The Collared Peccary (family Tayassuidae) is a New World relative of the pigs, and has a short tail and a piglike nose. Like the true pigs, it has sharp tusks with which it can defend itself if cornered. Peccaries travel in bands and are found only in the drier parts of the Southwest.

Deer The deer (family Cervidae), which range in size from the Moose to the tiny Key Deer of the Florida Keys, are easily recognized by their branching antlers, which are made of bone but are shed and replaced each year. In addition to the paired hoof marks of other hoofed animals, deer often leave smaller marks made by two dewclaws located behind the main toes. Although they will eat grass, deer are mainly browsers, cropping buds, leaves, and twigs. Some, like the Moose, are fond of aquatic plants, while the Caribou eats lichens during the severe Arctic winter.

Pronghorn The fleet-footed Pronghorn (family Antilocapridae), the world's fastest land mammal, is often called an antelope, but it belongs to a family of its own and is restricted to North America. Its curved horns have a bony core, as well as a horny covering that is shed and renewed each year.

Bison, Sheep, and Goats The cows, sheep, goats, and their relatives (family Bovidae) have permanent horns made of bone with a sheath that is also permanent. These horns are used both for defense and in combat between males during the mating season. They all live in herds, with males and females sometimes separated, and all are grazers rather than browsers like the deer.

ndex

[umbers in italics refer to ıammals mentioned as similar ɔecies.

Credits

Photographers
Roger W. Barbour (149, 171)
Sharon Cummings (23, 143)
Sharon Cummings / DEMBINSKY PHOTO ASSOC (155)
Jeff Foott (27, 33, 55, 89, 103, 115, 119, 121)
Carl Hanninen / PHOTO/NATS (69)
G. C. Kelley (19, 39)
Breck P. Kent (81, 109, 113, 135)
Dwight R. Kuhn (59, 147, 157, 165, 177)
Tom and Pat Leeson (45, 57, 123)
Gerard Lemmo (79)
Joe McDonald (49, 65, 97, 107, 131, 139, 145)
Tom McHugh / THE NATIONAL AUDUBON SOCIETY COLLECTION / PHOTO RESEARCHERS, INC. (167, 175)
Skip Moody / DEMBINSKY PHOTO ASSOC (21, 37)
C. Allan Morgan (51, 53, 71, 111, 151)
Alan G. Nelson / DEMBINSKY PHOTO ASSOC (35, 77, 95)
Stan Osolinski / DEMBINSKY PHOTO ASSOC (63)
James F. Parnell (125, 129, 133, 159, 161, 169, 173)
John Peslak (93, 153)
Rod Planck / DEMBINSKY PHOTO ASSOC (91, 117 ,163)
Len Rue, Jr. (47)
Leonard Lee Rue III (73)
Carl R. Sams / DEMBINSKY PHOTO ASSOC (25)
Larry Sansone (105)
Dick Scott / DEMBINSKY PHOTO ASSOC (127)
John Shaw (29, 41, 67, 83–87, 99, 101, 137, 141)
Alvin E. Staffan / THE NATIONAL AUDUBON SOCIETY COLLECTION/ PHOTO RESEARCHERS, INC. (61)
Robert L. Zakrison (31, 43, 75)

Illustrators
Tracks by Dot Barlowe, Dolores R. Santoliquido, and Sheila Ross
Other animal signs by Edward Lam

Cover photograph:
Coyote by Stan Osolinski / DEMBINSKY PHOTO ASSOC

Title page: Bighorn Sheep by Skip Moody / DEMBINSKY PHOTO ASSOCIATES

Spread (16–17): Eastern Cottontail by Sharon Cummings

Prepared and produced by Chanticleer Press, Inc.

Founding Publisher: Paul Steiner
Publisher: Andrew Stewart

Staff for this book:

Managing Editor: Barbara Sturman
Editor: Jane Mitzner Hoffman
Designer: Sheila Ross
Photo Editor: Timothy Allan
Production: Gretchen Baily Wohlgemuth
Editorial Assistant: Kate Jacobs

Original series design by Massimo Vignelli.

All editorial inquiries should be addressed to:
Chanticleer Press
665 Broadway, Suite 1001
New York, NY 10012

To purchase this book or other National Audubon Society illustrated nature books, please contact:
Alfred A. Knopf, Inc.
1745 Broadway
New York, NY 10019
(800) 733-3000
www.randomhouse.com

NATIONAL AUDUBON SOCIETY

The mission of NATIONAL AUDUBON SOCIETY *is to conserve and restore natural ecosystems, focusing on birds, other wildlife, and their habitats for the benefit of humanity and the earth's biological diversity.*

One of the largest environmental organizations, AUDUBON has 550,000 members, 100 sanctuaries and nature centers, and 508 chapters in the Americas, plus a professional staff of scientists, educators, and policy analysts.

The award-winning *Audubon* magazine, sent to all members, carries outstanding articles and color photography on wildlife, nature, the environment, and conservation. Audubon also publishes *Audubon Adventures*, a children's newspaper reaching 450,000 students. Audubon offers nature education for teachers, families, and children through ecology camps and workshops in Maine, Connecticut, and Wyoming, plus unique, traveling undergraduate and graduate degree programs through *Audubon Expedition Institute*.

AUDUBON sponsors books, on-line nature activities, and travel programs to exotic places like Antarctica, Africa, Baja California, and the Galápagos Islands. For information about how to become an Audubon member, to subscribe to *Audubon Adventures*, or to learn more about our camps and workshops, please contact:

NATIONAL AUDUBON SOCIETY
Membership Dept.
700 Broadway, New York, NY 10003-9562
(800) 274-4201 or (212) 979-3000
http://www.audubon.org/